Carbon sequestration in tropical grassland ecosytems

Carbon sequestration

in tropical grassland ecosytems

edited by:

L. 't Mannetje

M.C. Amézquita

P. Buurman

M.A. Ibrahim

Wageningen Academic
P u b l i s h e r s

ISBN 978-90-8686-026-5

First published, 2008

Wageningen Academic Publishers
The Netherlands, 2008

Preface

This book shows and substantiates that well managed, sustainable, productive grassland and silvopastoral ecosystems in tropical America are capable of sequestering and storing large amounts of Carbon (C) in the soil, comparable with native forests.

Results are reported of a large five year on-farm research project carried out in tropical America (Colombia, Costa Rica). Soil and vegetation C stocks of long-established and experimental pasture, forage bank and silvopastoral systems on commercial farms were compared with those of adjacent forest and degraded land. The objective was to identify production systems that both increase livestock productivity and farm income and, at the same time, contribute to a reduction of CO_2 accumulation in the atmosphere, which is accepted by many politicians and scientists to be the overriding cause of global warming.

The project was carried out in the Andean Hillsides of the semi-evergreen forest in Colombia, the Colombian humid Amazonian tropical forest ecosystem, the sub-humid tropical forest ecosystem on the Pacific Coast of Costa Rica and the humid tropical forest ecosystem on the Atlantic Coast of Costa Rica.

Apart from data collection and analysis, there are also chapters on data quality in relation to soil variability and data consistency, extrapolation of results by means of modelling and a study to identify regions in other parts of tropical America with similar environmental conditions as exist in the four ecosystems of this study, where the results of this study might be applicable.

This was a multinational project carried out by teams of scientists of Universities and Research Institutes of Colombia, Costa Rica and The Netherlands. The project also contributed to the training of graduate students and junior research staff in the participating organisations.

Table of contents

Executive summary

Chapter 1. Introduction

The Dutch government funded a project that evaluated improved grassland and silvopastoral systems in Tropical America (Colombia and Costa Rica) in terms of C sequestration and farm benefits with the objective to reduce CO_2 accumulation in the atmosphere and to eleviate poverty. Research teams of the Centre for Research on Sustainable Agricultural Production Systems (CIPAV), Cali, Colombia, the University of Amazonia, Florencia, Caquetá, Colombia, the Tropical Agricultural Research and Higher Education Center (CATIE), Turrialba, Costa Rica, the International Center for Tropical Agriculture (CIAT), Cali, Colombia and Wageningen University and Research Centre (WUR), Wageningen, The Netherlands carried out the project from 2001-2006. The project was carried out in Colombia and Costa Rica in the following ecosystems:

- eroded Andean Hillsides of the semi-evergreen seasonal forest ecosystem (Colombia);
- the Amazonian humid tropical forest ecosystem. (Colombia);
- sub-humid tropical forest ecosystem on the Pacific Coast (Costa Rica);
- the humid tropical forest ecosystem in the Atlantic Coast (Costa Rica).

The objectives of the project were to:
1. Estimate soil and vegetation C stocks of long established (10-20 years) pasture and silvopastoral land use systems and compare these with stocks under native forest and degraded grasslands.
2. Estimate C sequestration rates of newly established improved pasture and silvopastoral land use systems on degraded land.
3. Estimate the socio-economic benefits to farmers of establishing improved pasture, agropastoral or silvopastoral land use systems in degraded areas.
4. Identify, within each ecosystem, land use systems that are economically attractive to the farmer, help to alleviate poverty and have a high capacity for C storage.
5. Extrapolate project results to similar environments in Tropical America.
6. Provide recommendations at local, national and international level, regarding policy decisions to mitigate and adapt to the adverse effects of climate change, taking into account appropriate land use that provides environmental and socio-economic benefits to farmer population.

Chapter 2. Methodology of bio-physical research

The long-established land use types studied for C stocks were native forest, unimproved grass pastures, improved grass-legume pastures, degraded pastures, silvopastoral systems and forage banks. Improved land use types were established on degraded pastures, consisting of monocultures of improved grasses, grass-legume pastures and multi-species forage banks. The effect of these improved systems was evaluated after about three years. The studies were carried out on small to large private farms and on two research farms. Soil and vegetation samples were collected and analysed to estimate C contents.

Chapter 3. C stocks and sequestration

The following conclusions can be drawn:

Long-established land use systems
- Mean soil C in native forests and in long-established improved pastures of all ecosystems studied were about the same (157 vs. 160 t/ha). However, mean total C (in soil and above ground biomass) in forest was 40% more than in grasslands (261 vs. 162 t/ha).
- In pasture and silvopastoral systems, more than 80% of the total C in the ecosystem was in the soil. Therefore, even small increases of the soil-C stock in such systems contribute significantly to C sequestration.
- Land use was the main factor explaining changes in C stocks.
- The results of this and other studies showed that forests store large amounts of C in the tree biomass compared to that sequestered in the soil. Storing C in trunks of trees may represent one way of increasing permanent C stocks under sustainable harvesting and processing of timber.
- Highly productive agroforestry systems, including silvopastoral systems, can play an important role in C sequestration in soils and in the woody biomass. Well managed silvopastoral systems can improve overall productivity, while sequestering C, which is a potential additional economic benefit for livestock farmers.

Short-term (3-3.4 years) experiments
- Newly established improved pastures and natural forest regeneration sequestered on average in all ecosystems 6 t C/ha/yr in the soil during 3.4 years.

- The land use systems of improved grasses and grass-legume mixtures and natural regeneration of degraded pasture were most effective in sequestering C in the soil, with yearly increments varying between 2 and 9 tons of C per hectare.

Chapter 4. Analysis of soil variability and data consistency

Homogeneity of soils of experimental plots to warrant comparison of the effects of land use systems was tested using logical relations between soil properties. Similarly, consistency of analytical data was checked and inconsistent information was discarded.

Chapter 5. Factors affecting soil C stocks: a multivariate analysis approach

Because differences in C stocks are not solely dependent on differences in land use, but also on internal soil properties, relations between measured soil properties and C stocks were studied using factorial analysis. The results show that C stocks are always negatively correlated with soil bulk density (BD), which is probably an effect of more intense rooting in improved systems. The factorial analysis also allowed identification of physical differences between experimental plots.

Chapters 6 and 7. Socio-economic methods and results

Benefit-cost analyses were carried out to evaluate the financial feasibility of investing in different land use management systems with capacity to sequester C in typical livestock farms located in the Andean Hillsides and the Amazon Region of Colombia and the Sub-Humid Tropical Forest of Costa Rica. Under the model's assumptions, investments in adopting improved pastures with trees and in fodder banks showed positive incremental net present values (NPV), with internal rates of returns to the farmer's own resources between 19% and 29%. All models showed a likelihood of obtaining positive NPVs in more than 90% of trials, indicating that the investments were almost risk-free. Applying a payment for C had a marginal effect on model results.

The size of the investment for the establishment of land use management systems with capacity to sequester C varied considerably among the three study sites. In the case of the Colombian Amazon, the investment associated with the establishment of one hectare of improved pastures associated with legumes and trees amounted to US$ 435 and for the same association plus 0.3 ha of fodder bank was US$ 512. Investment costs in the Andean Hillsides of Colombia for the same land use management systems were US$ 417 and 1272, respectively. In the case of Costa Rica, the cost of establishing 1 ha of improved pastures associated with legumes and trees (under natural regeneration) was US$ 161, while the cost of establishing 0.9 ha of fodder bank was US$ 660.

Chapter 8. Reflections on modelling and extrapolation in tropical soil C sequestration

Various approaches of modelling and extrapolation in tropical soil C sequestration are discussed. Among others, it is argued that an approach based on statistical methodology is to be preferred to the commonly used Process Based Simulation modelling techniques. Application of the results leads to underpinning and justification of the extrapolation procedure as carried out in the C-sequestration project described extensively in this book. Apart from the more traditional techniques, a methodology called 'consecutive chronosequence' is proposed as a promising alternative approach to deal with the important question of the identification of the 'best' Land Management System with respect to soil C sequestration potential.

Chapter 9. Extrapolation of results to similar environments in tropical America

Extrapolation of results:
- *Ecosystem 1: Andean Hillsides, Colombia.* An area of about 7,000 km^2 has *Umbric Andosols* and 3,000 km^2 *Dystric Cambisols* with identical Length of Growing Period (LGP) and slope/elevation conditions as those observed in ecosystem 1.
- *Ecosystem 2: Humid Tropical Forest, Amazonia, Colombia.* This ecosystem is widespread in the undulating and flat lowlands of the upper Amazon. Soils in these areas are mainly *Haplic Acrisols* and *Haplic Ferralsols*. Some 288,000 km^2 in Brazil, Colombia, Ecuador, Gyana, Peru and Venezuela are *Haplic Acrisols* and 80,000 km^2 *Haplic Ferralsols* with similar LGP and slope/ elevation conditions (flat topography) as ecosystem 2. Sloping topography occurs in minor areas in ecosystem 2 (Figures 7 and 8).
- *Ecosystem 3:- Humid Tropical Forest, Atlantic Coast, Costa Rica.* An area of less than 750 km^2 in coastal Rica, Guatemala and Panama, has *Dystric Cambisols* with identical LGP and slope/elevation conditions as ecosystem 3.

- *Ecosystem 4:- Sub-humid Tropical Forest, Pacific Coast, Costa Rica.* An area of about 1,300 km^2 in Costa Rica, Nicaragua and Panama has *umbric, dystric* or *eutric Cambisols* with identical LGP and slope/elevation conditions as ecosystem 4.

Uncertainties in these extrapolations, due to different systems of soil classification, small scales of available soil maps, approximations of topographical unitsand complexity of soil mapping units, are discussed in Chapter 9.

Chapter 10. Conclusions and policy recommendations

Results from this project are significant regarding the positions to be adopted by Tropical American countries *vis-à-vis* the United Nations Conference on Climate Change and the negotiations carried out through its implementing bodies. The development of a strategy to include improved pasture and silvopastoral systems as well as the conservation of natural forests (avoided deforestation) is required as eligible projects under the Clean Development Mechanism for the next implementation phase of the Kyoto Protocol. In the case of improved pastoral and silvopastoral systems, this project provides ample evidence that those systems present a valid alternative to increase C accumulation, particularly in soils. In the case of native forests, the results from this investigation underline, once again, the importance native forests play as C stocks compared to other land uses. This fact becomes particularly relevant given the high deforestation rates the region continues to suffer.

Findings from this project are robust enough to encourage governments in Tropical America to review their policies that favour reforestation and afforestation as a means to recover degraded pastures and, simultaneously, to mitigate climate change. This study clearly indicated the capacity of improved pasture and silvopastoral systems to recuperate degraded areas and, at the same time, to provide an attractive economic alternative to farmers.

The decision on the best alternative (reforestation, afforestation, improved pasture and silvopastoral systems) is, in the end, site specific and must include environmental as well as social considerations, such as norms and regulations on water management, soil and biodiversity protection, poverty levels, labour demands, capital requirements and cultural traditions, among others.

Foreword: in search of new horizons in socio-environmental policies

Manuel Rodríguez-Becerra

The results of the research project as presented in this book can contribute in a significant way to advances in the science related to global warming and the appropriate land use of Latin American tropical (LAT) ecosystems. It can also play an important role in the development of new public policies both at the regional and global level.

The research carried out offers an assessment of C accumulation in pasture and silvopastoral systems in comparison with native forest and degraded pasture in three LAT ecosystems: tropical humid forest, sub-humid tropical forest and the Andean Hillsides in Colombia and Costa Rica. In addition, the research results highlight how improved systems are one of the possible strategies that can be used to alleviate poverty of families working on small and medium sized farms.

The conclusions and results obtained from the research can contribute to formulating public policy in three key plots: (1) the research offers new views and approaches that can play an important role in international negotiations leading up to the later phases of the Kyoto Protocol. In particular, the research results obtained can be used to include new eligible projects regarding land use within the Clean Development Mechanism (CDM) from 2012 onwards, the period which also marks the end of the first commitment period of the Kyoto Protocol; (2) the research results offer governments of LAT countries new ways to formulate public policy aimed at helping developing countries adhere to the specific agreed aims of the UN Framework Convention on Climate Change and the Kyoto Protocol; and (3) the research results can contribute to the recovery of degraded soil in LAT countries which is one of the main socio-environmental challenges facing the region. In fact, measures aimed at recovering degraded soil are inextricably linked with ways used to combat poverty among those farmers most vulnerable. In this way this research can contribute as well to the fulfillment of the United Nations Millennium Development Goals and of the objectives of the UN Convention to Combat Desertification.

It is no mere accident that the research carried out is particularly relevant in the plot of scientific research and formulating public policy both at the national and global level. From the outset, the research which is multinational in nature, was conceived and designed so that it could be used within a sustainable development context. Furthermore, it is important to highlight that as a result of the research carried out, developing countries, in particular LAT countries, are now able to play a more important role in international negotiations about the environment. In practice, as has been the case in the past, the role played by the majority of developing countries during international negotiations and agreements about the environment has been marginal as a result of the lack of scientific data obtained in their own countries.

The need for action and policy makers

Climate change is a reality that many leading politicians and the business community prefer to ignore or refuse to recognise. The fact that the US administration did not ratify the Kyoto Protocol is just but one example. Such a stance is also adopted and supported by powerful business lobby groups who continue to manipulate the media so that the public remain ignorant of scientific evidence which points to global warming and that human activity is the main cause of this phenomenon.

Indeed the American Petroleum Institute (Krugman, 2006) pursues a strategy, as outlined in an internal document leaked to the public in 1998 that reveals the institute providing, 'moral and logistical support' to those people who still refuse to accept that climate change exists. The strategy aims to 'raise question marks about climate change' and attempts to discredit 'prevailing scientific knowledge'.

Such a strategy has been very effective, 'A 2003 study, by Maxwell Boykoff and Jules Boykoff, of reporting on global warming in major newspapers found that a majority of reports gave the skeptics – a few dozen people, many if not most receiving direct or indirect financial support from Exxon Mobil – roughly the same amount of attention as the scientific consensus, supported by thousands of independent researchers' (Krugman, 2006).

Fortunately not all the big companies who produce fossil fuels sponsor such types of campaigns, as in the case of BP and Royal Dutch Shell. But the influence of those who still refuse to accept prevailing beliefs about climate change has produced negative effects. It seems a similar pattern to the pathetic case of the tobacco industry which for almost two decades managed to successfully discredit scientific evidence about the lethal effects of cigarette smoking, which had been known since the 1930's.

However, there is scientific evidence, which is becoming more widespread and aggressive in nature which highlights that certain human activities are bringing about climate change. Such research raises the serious extent of the problem and the possible consequences of such human activities across different geographical regions and timescales. But as has been suggested above, despite increasing scientific evidence about the existence of climate change, such developments and trend continue.

In addition, those countries who ratified the Kyoto Protocol, as well as those public organisations and scientists who are genuinely committed to combating this global threat must do all that is possible to spread scientific research about climate change among diverse sectors of the public. This can be one of many strategies used to promote greater political will both within countries and internationally.

Recent events provide some hope on the possibility of entering on the right track. New reports and findings emphasise the need of taking strong action now. The Stern Review on the Economics of Climate Change 'estimates that if we don't act, the overall costs and risks of climate change will be equivalent to losing at least 5% of global Gross Domestic Product (GDP) each year, now and forever. If a wider range of risks and impacts is taken into account, the estimates of damage could rise to 20% of GDP or more. In contrast, the cost of action – reducing gas emissions to avoid the worst impacts of climate change – can be limited to around 1% of global GDP each year' (2006). More recently, 'Climate Change 2007', also known as the Fourth Assessment Report of the Intergovernmental Panel on Climate Change (IPCC), seems to have had a high impact on the public opinion and the political and business realm.

Various regional and international leaders, such as Tony Blair (2004) and Al Gore (2006), have shown that they are increasingly committed to strengthening national, regional and global policies regarding the environment and climate change, which is an urgent need. They are joined by imminent individuals who promote scientific research and help to make it popular, including Flannery (2006), Hansen (2006) and Kolbert (2006). The development of scientific research about climate change and ways in which to prevent and mitigate it, as well as how to better adapt vulnerable ecosystems to lessen their negative impacts, are key factors in the creation of a more robust legal framework, norms and policies.

Research findings and CDM

Within the CDM, reforestation and afforestation projects are being included as eligible projects for the first commitment period of the Kyoto Protocol (2008-2012). Offering financial incentives to promote reforestation and afforestation projects in developing countries is a very positive step. However, the CDM does not include conservation or deforestation avoidance as eligible projects, something which many consider to be a negative aspect of the CDM.

It is clear that the destruction of natural forests in LAT ecosystems, as in other regions, continues. For this reason, it is difficult to understand why the CDM does not promote financial incentives which encourage the conservation of natural forests in LAT ecosystems. The research results as presented in this book support the view that emphasises the importance of natural forests as C 'sinks' when compared with other land uses. This research project, which examined specific ecosystems over a five-year period, concluded that, in terms of C accumulated in the whole ecosystem *(soil + biomass)*, native forests show the highest levels of C stocks, followed by improved pasture and silvopastoral systems (Amézquita *et al.*, 2004, 2005a,b,c).

As shown in this research project, natural forests should be eligible as part of the CDM, which is an issue that has yet to be decided upon within the Kyoto Protocol. Furthermore, natural forest projects should be part of the CDM particularly when considering the innumerable environmental and social benefits that natural forests can bring.

In addition to the conclusions as discussed above, which are backed by diverse studies and governments during international negotiations, it is important to highlight that the new findings obtained about the roles played by different forest ecosystems in the accumulation of C in soil is one of the most significant conclusions raised, not only in the field of science but when formulating public policy. One of the main findings was that pasture and silvopastoral systems can accumulate between 8 to 80 per cent more C in the soil than the prevalent degraded pastures. In addition, in terms of the amount of C accumulated in the soil, depending on the climatic and environmental conditions of a particular plot, improved and well-managed pasture and silvopastoral systems store C levels similar to and even greater than those found in native forests (Amézquita *et al.*, 2004, 2005a,b,c).

These findings highlight that the future of degraded pastures in LAT ecosystems does not necessarily involve reforestation or afforestation projects, an approach which is basically being promoted by the CDM. Furthermore, the research indicates that there is little scientific founding to support the increasingly generalised view which prioritise reforestation and/or forestation as alternatives to recover degraded soil and at the same time help mitigate climate change. Furthermore, the research indicates that there is little scientific founding to support the increasingly generalised view which features reforestation and or forestation as the best alternatives to recover degraded soil and at the same time help mitigate climate change.

To summarise, improved pasture and silvopastoral ecosystems are valid ways with which to increase the accumulation of C in the soil. In other words, such measures need to be included as eligible projects within the new phases of the CDM.

Socio-economic relevance: soil restoration and poverty alleviation

The project concludes that such improved systems can also play an important role in providing solutions, in both the short and medium term, to the socio-economic problems facing farmers who belong to the poorest groups in the region, particularly those families who own small and medium sized farms on degraded soil.

The incomes of families who own farms in improved systems are substantially higher than those families who own farms on degraded soil. This is why it is important to compare ways to establish improved pasture and silvopastoral systems on degraded soil with other alternatives such as reforestation and afforestation on the same degraded soil. As emphasised earlier, forests accumulate a greater incremental amount of C than other terrestrial systems. Indeed, forests provide other environmental benefits but as the research shows establishing forests on degraded soils may not be the best approach to tackle and improve the basic needs among the poorest groups in society.

The relevance of the research results obtained are better understood if one takes into account the fact that in LAT countries there are 318 million hectares of degraded soil, of which a large proportion is on small and medium sized farms. Without doubt, degraded soil is one of the greatest social and environmental problems facing the region, but one which has received little attention among governments. In many cases, the issue of degraded soil is rarely a social and or environmental concern and focus in many countries (Rodríguez Becerra and Espinoza, 2002; Bucher *et al.*, 2000).

To summarise, making policy decisions about which measures to adopt in order to tackle soil degradation in the LAT region, which may help mitigate climate change, must also take into account how such measures may help to eradicate poverty. In addition to how this research may influence policy implemented to mitigate climatic change at the global, regional and local level, it is important to look at the problem taking into account its many different environmental considerations. These include C capture and its impact on biodiversity protection, water control and protecting against erosion.

Future research

There is a need to examine the fundamental issues of hydrological externalities in diverse systems such as pasture, silvopastoral and improved agroforestry ecosystems and comparing them with hydrological externalities found in degraded pastures, natural forests and planted forests.

Today, we are aware that reforestation does not always have positive hydrological effects (Kaimowitz, 2000). This assertion opposes the widely accepted view that claims that reforestation of degraded soil is the magic formula with which to recover water sources, both in terms of quality and quantity, as well as being a way to restore the regularity of water cycles and decrease erosion. Such ideas have basically evolved into myths, which have been backed by hundreds of policies and reforestation programs adopted in the region. For this reason, it is so important that future research should aim to find out to what extent establishing pasture, silvopastoral and agroforestry systems on degraded soil can contribute to the protection of water basins.

Finally, public policy decision makers, both national and international, should take into account the results of the research as outlined in this book. Furthermore, it is the policy makers who have a great responsibility to continue supporting further research, which will bring new ideas and approaches that can be used to combat the most important environmental challenges we face, including global warming, soil erosion and the depletion of water resources, while also helping to tackle poverty among the most vulnerable farmers living in the LAT ecosystems.

Chapter 1. Introduction

M.C. Amézquita, E. Murgeitio, B.L. Ramirez and M.A. Ibrahim

This book reports on research into carbon (C) storage and sequestration in pasture and silvopastoral systems, with reference to native forest and degraded land, in four ecosystems in tropical America: Andean Hillsides, Colombia; Amazonian humid tropical forest, Colombia; humid tropical forest, Atlantic coast, Costa Rica; and sub-humid tropical forest, Pacific coast, Costa Rica. Research was carried out by the project *'Research Network for the Evaluation of C sequestration Capacity of Pasture, Agropastoral and Silvopastoral Systems in the American Tropical Forest Ecosystem'*, financed by The Netherlands Cooperation as Activity CO-010402 and carried out by research teams of the Centre for Research on Sustainable Agricultural Production Systems (CIPAV), Cali, Colombia, the University of Amazonia, Florencia, Caquetá, Colombia, Cali, Colombia, the Tropical Agricultural Research and Higher Education Center (CATIE), Turrialba, Costa Rica, the International Center for Tropical Agriculture (CIAT), Cali, Colombia and Wageningen University and Research Centre (WUR), Wageningen, The Netherlands. The project responds to the United Nations Framework Convention on Climate Change (UNFCCC), Kyoto Protocol, the Bonn Agreement (UNFCCC COP6, 2001), the Marrakech Conference (UNFCCC COP7, 2001) and The Netherlands Implementation of Clean Development Mechanism (CDM) and related research on adaptation alternatives (October 22, 2001). It consulted the Intergovernmental Panel on Climate Change Guidelines (IPCC, 1999). It also responds to national commitments of Colombia and Costa Rica to the UNFCCC related to mitigation and adaptation strategies and activities to protect vulnerable ecosystems in their countries and to national policies for environmental protection and poverty reduction. The project lasted five years (2001-2006).

Tropical America: land use, land use change, economic and environmental importance of pasture and silvopastoral production systems

Tropical America comprises Mexico, Central America, the Caribbean and South America, excluding Argentina, Chile and Uruguay. It covers 11% of the world's land area in which 8% of the world's population lives. Twenty three percent of tropical America's population are farmers (FAO, 2002).

The region is strongly associated with deforestation of native forests in order to cultivate agricultural crops, pastures and silvopastoral systems for animal production. This represents the most important change in land use in tropical America in the last 50 years (Binswanger, 1989; Kaimowitz, 1996; Pfaff *et al.*, 2004). The driving forces of deforestation are poverty in the small-farmers' sector, government policies and population pressure.

In many tropical American countries, most deforestation results from the actions of poor subsistence farmers. Forest land transformation started with slash and burn practices to establish food crops, such as maize, cereals and fruits. A loss in soil fertility and production associated with poor management resulted in severe soil degradation. At the end of this cycle, pasture systems were established, which, under inappropriate management, degraded the soil even more, leaving severely degraded land with pastures of low quality and productivity, with increased risk of erosion.

Some degraded pasture land has been recovered through the establishment of improved pastures consisting of selected grasses and legumes and of agropastoral and silvopastoral systems. However, the recovery of degraded pasture land, even if supported by government incentives or loans, is adopted by the farmer only if it represents an attractive economical option.

Pastures, agropastoral and silvopastoral systems in tropical America cover 77% of its 548 million ha of agricultural land and 11% of the world's agricultural land (FAO, 2002). The most important ecosystems of tropical America for pasture and silvopastoral production are the savanna (250 million ha) and tropical forest (44 million ha).

Semi-intensive grazing systems of beef and dual-purpose cattle and other ruminants are predominant in the tropical forest ecosystem, accounting for 78% of the meat and 41% of the milk produced in tropical America (Lascano *et al.*, 2001). Although there is a large area of pasture, agropastoral and silvopastoral land available, production levels are low due to degraded or low-quality pastures and non-sustainable management practices. It is estimated that more than 40% of pastures are severely degraded in Central America (Szott *et al.*, 2000) and even more so in the rest of tropical America (CIAT, 1999). Land use that exposes the soil to direct action of rain or wind will accelerate erosion. One of the principal causes of soil erosion is cattle raising on slopes exceeding 30%. As a result, livestock keeping on slopes is considered prejudicial for the conservation of the environment. However, appropriate pasture and animal management will reduce the erosion rate to levels such as those observed on forested slopes (IVITA, 1981).

Both in the savanna and tropical forest ecosystems, pasture systems are important land use systems for animal production, poverty alleviation, and the generation of environmental services, in particular C sequestration and erosion control. The effect of improved pasture systems and management practices on increases in milk and meat production and human welfare has been amply documented (CIAT, 1976-1999). However, their effect on increases in C sequestration and the cost of degradation in terms of C losses has not yet been extensively documented.

Research on C sequestration in pasture systems in the Savanna ecosystem of tropical America was initiated by CIAT (Fisher *et al.*, 1994), who demonstrated the potential of soil C accumulation of deep-rooted grasses (*Andropogon gayanus* and *Brachiaria humidicola)* in monoculture and associated with forage legumes (*Arachis pintoi* and *Stylosanthes capitata*). Ibrahim *et al.* (2001) showed that annual increments of soil organic C in pure *B. humidicola* pastures was 7 t/ha/yr and 11 t/ha/yr in a silvopastoral system with *Acacia mangium.*

The main objective of the present research project was to contribute to sustainable development, poverty alleviation and mitigation of the undesirable effects of climate change, by evaluating the socio-economic benefits and the capacity of C sequestration of a range of pasture, agropastoral and silvopastoral ecosystems, in small, medium-size and large farms in vulnerable ecosystems of the American tropics.

Project specific objectives were:
1. Estimate soil and vegetation C stocks of long-established (10-20 years) pasture and silvopastoral land use systems comparing them with those from adjacent native forest and degraded land.
2. Estimate C sequestration rates of newly-established improved pasture and silvopastoral systems on degraded land, through short-term replicated small plot experiments.
3. Estimate the socio-economic benefits to farmers of establishing improved pasture, agropastoral or silvopastoral land use systems in degraded areas.
4. Identify, within each ecosystem, land use systems that are economically attractive to the farmer, help to alleviate poverty and have a high capacity for C storage.

5. Extrapolate project results to similar environments in tropical America.
6. Provide recommendations at local, national and international level, regarding policy decisions to mitigate and adapt to the adverse effects of climate change, taking into account appropriate land use that provides environmental and socio-economic benefits to farmer population.

Description of ecosystems

The Andean Hillsides ecosystem

The Andean Hillsides ecosystem covers 96 million ha in Perú, Ecuador, Colombia and Venezuela (Pachico *et al.*, 1994). In Colombia, 70% of the country's total population lives in this region, with 78 inhabitants/km^2 compared to 29 inhabitants/km^2 in the rest of the country (DANE, 1996). Pasture, agropastoral and silvopastoral systems have important socio-economic significance in this region, as they produce milk and meat. Improved and well-managed pasture systems contribute to poverty reduction, recovery of degraded areas and delivery of environmental services, including C sequestration and erosion control.

Field research was carried out on small farms that use grazing and cut-and-carry animal production systems in El Dovio and Dagua, at 1,200-2,000 m.a.s.l., with acid soils (pH 5.2-6.2), mean temperature 14.0-20.0 °C during the growing season, mean rainfall 1,500-1,900 mm/yr exceeding 60% of potential annual evapo-transpiration during 6-9 months and moderate to steep slopes (15-63% depending on land use system).

The Amazonian humid tropical forest ecosystem

The importance of the Amazonian humid tropical forest ecosystem lies not only in its great surface area and the huge amount of renewable resources they represent, but also in their environmental, social and economical properties. Of the *ca.* 300 million ha of the Amazon basin, 75% of soils (oxisols and ultisols) are characterised by (1) low fertility, (2) can keep fertility only through the incorporation of organic matter, (3) acidity with a high Al content, (4) very compactable, (5) very erodable, (6) low in cation exchange capacity, and (7) low in natural P content (TCA, 1991). The ecosystem is characterised by a mean annual rainfall of 3,800 mm, a heavy rainy season between March and October, less than 2,000 hrs/yr of sun shine, an average mean temperature of 23 °C, and relative humidity of 83%. These are significant constraints for crop and livestock production (Sánchez *et al.*, 1982). In addition, both ignorance about the ecological relationships and inappropriate land management, have led to a great loss of biodiversity.

The general types of land use in the area are livestock production, cropping systems, and forestry (CADMA, 1992). Livestock production is the most important economic activity in this ecosystem, with 60% of the population depending on it. According to FAO (2000) 23% of the agricultural land in the Amazon is used for extensive pasture production under grazing, both for beef and dual-purpose cattle production (meat and milk). Dual-purpose livestock production is carried out mostly by small farmers, with low production per unit area. The very fast loss of soil fertility occurring after deforestation leads to a large percentage of the land being abandoned each year. This process is common in the three main settling areas of the Colombian Amazon, namely Caquetá, Guaviare and Putumayo, as well as in other areas in Brazil, Peru and Ecuador. The expansion of extensive cattle production with inappropriate management in the Amazon carries with it a high degradation risk (Fearnside, 1979; Sioli, 1980). Serrao *et al.* (1978), quoted by Hecht (1992), estimated that almost half of the 500,000 ha dedicated to pastures in the Brazilian Amazon were in an advanced stage of degradation. The livestock enterprises in the Amazon need to improve and increase the sustainability of animal production, as well as the social and economic conditions of the people who depend on it.

The present research project has worked on five large and medium-sized farms under grazing in the humid tropical Amazon piedmont in the Caquetá Department with flat and mildly sloping topography with a mean altitude of 270 m.a.s.l. Three were commercial dual-purpose cattle farms and two experimental farms.

Sub-humid and humid tropical forest ecosystems, Costa Rica

Study sites in Costa Rica were selected in two contrasting ecological regions: Esparza, located in the sub-humid seasonally dry hillsides along the Pacific Coast, and Pocora, located in the humid lowlands along the North Atlantic Coast, close to Guapiles. These regions differ mainly in the quantity and distribution of rainfall during the year (Esparza 2,500 and Pocora 4,500 mm). The soils in Esparza (10-1,400 m.a.s.l) are mainly afisols, inceptisols and in Pocora (200 m m.a.s.l) inceptisols. Cattle production systems in the Esparza region are dual-purpose and beef fattening systems based on grazing of *Hyparrhenia rufa* and brachiaria pastures. In the Pocora region they are suckling cow and calf and dual-purpose systems based on pastures of *Ischaemum ciliare and Brachiaria* spp. as the main source of feed.

The traditional silvopastoral systems in the sub-humid zone at Esparza at the Pacific Region of Costa Rica developed as a result of natural regeneration of different tree species in pastures. Trees are present in irregular patterns either isolated or growing in small clusters. These systems contained different tree species, the most abundant being *Cordia alliodora, Cedrella odorata, Enterelobium ciclocarpum, Tabebuia rosea, Acrocomia aculeata* and *Psidium guajava*. Tree densities in the traditional silvopastoral system at Esparza varied between 30 and 51 adult trees/ha. There were different stages (e.g. saplings, juvenile and adult) of regeneration of the tree species (C). In the tropical humid ecosystem at Pocora in the Atlantic Region of Costa Rica an alley-pasture silvopastoral system was established in 1993 with the multi-purpose tree *Acacia mangium*, planted at 3 m distances within rows that were 8 m apart. Between rows the herbaceous legume *A. pintoi*, was planted.

Chapter 2. Methodology of bio-physical research

M.C. Amézquita, M. Chacón, T. LLanderal, M.A. Ibrahim, J. Rojas and P. Buurman

Introduction

The bio-physical research common across ecosystems and land use systems, included:
a. Quantification of C stocks of a range of long-established commercial land use systems that were compared with the reference systems degraded pasture and native forest.
b. Establishment of small-plot experiments on degraded pasture sites, to quantify and compare after 3.4 years the level of C accumulation of improved pasture and silvopastoral systems with degraded pasture.

C stocks in long-established land use systems

The land use systems for the Andean Hillsides are presented in Table 1, for Amazonia in Table 2 and for Costa Rica in Table 3.

In Costa Rica the project studied simple forage banks consisting of a single tree or shrub species (e.g. *Leucaena leucocephala* or *Cratylia argentea*) at high densities between 15,000 and 30,000 plants/ha) and multi-strata forage banks consisting of two to four species of leguminous and non-leguminous trees and shrubs. In Esparza, a *C. argentea* forage bank was established in 1994 at a density of 20,000 plants/ha. This forage bank was managed in a cut-and-carry system harvested every four months. The secondary forest had developed from a fallow of degraded pasture to a secondary succession of forest without grazing. The secondary forest sites in Esparza were 9-12 years old at the time of sampling and had a species composition that was typical of the forest in the sub-humid tropics.

Table 1. Land use systems - Andean Hillsides, Colombia.

El Dovio		Dagua	
Land use	Age (years)	Land use	Age (years)
Native forest	26	Native forest	40
B. decumbens pasture	16	B. decumbens pasture	16
Forage bank[1]	15	Natural regeneration of a	40
Degraded pasture:	16	degraded Hyparrhenia rufa	
Pennisetum purpureum, c.v.		pasture	
King grass		Forage bank[2]	14
		Degraded soil	40

[1]Trichanthera gigantea, Morus spp., Erythrina edulis, Boehmeria nivea and Tithonia diversifolia.
[2]Trichanthera gigantea, Morus spp., Erythrina fusca and Tithonia diversifolia.

Table 2. Land use systems Amazonia, Colombia.

Flat topography[1] 'La Guajira', 'Santo Domingo' and 'Cartagena del Chairá' farms		Sloping topography[2] 'Pekín' and 'Balcanes' farms	
Land use	Age (years)	Land use	Age (years)
Native forest	>100	Native forest	>80
Native forest	50		
B. decumbens pasture	10	B. decumbens pasture	15
B.humidicola pasture	10	B. humidicola pasture	15
B. decumbens + native legume pasture	10	B. decumbens + native legumes pasture	15
B. humidicola + native legume pasture	10	B. humidicola + native legume pasture	15
Degraded pasture (Paspalum notatum)	40	Degraded pasture (Paspalum notatum)	40

[1]0–5% slope; [2]7–15% slope.

Table 3. Land use systems in Costa Rica.

Esparza, sub-humid tropical forest		Pocora, humid tropical forest	
Land use	Age (years)	Land use	Age (years)
Native forest	>50	Native forest	>50
Secondary forest	11	*Ischaemun ciliare* pasture	>70
B. decumbens pasture	8	*B. brizantha* pasture	19
Hyparrhenia rufa pasture	>100	*B. brizantha + A. pintoi* pasture	16
Silvopastoral system: *B. brizantha + Cordia alliodora + Guazuma ulmifolia*	9	Silvopastoral system: *Acacia mangium + A. pintoi*	15
Forage bank: *Leucaena* and *Cratylia argentea*	13	Degraded pasture	>70
Hyparrhenia rufa – weedy pasture	>50		

Research sites and farms

C stock measurements were carried out on farms described in Table 4 for the Andean Hillsides, in Table 5 for Amazonia and in Table 6 for Costa Rica.

Selected farms with flat topography in Amazonia were:
- 'Santo Domingo' a poorly managed experimental farm belonging to the Universidad de la Amazonia (60 ha) with degraded pasture (Table 5).
- 'Guajira' a well managed commercial livestock farm (250 ha) with partially intervened native forest and improved pasture.
- 'Cartagena del Chairá' farm with 200 ha non-intervened native forest.

Selected farms located on sloping topography were:
- 'Balcanes', a poorly managed experimental farm (60 ha) belonging to Universidad de la Amazonia (degraded pasture) (Table 5).
- 'Pekín' a well-managed commercial livestock farm (800 ha) with partially intervened native forest and improved pasture systems.

These five farms are located near Florencia, Caquetá.

Table 4. Experimental farms at El Dovio and Dagua, Andean Hillsides, Colombia.

Farm name	Area (ha)	Latitude and longitude	Altitude (m.a.s.l.)	Rainfall (mm/year)	Land class[1]	Soil type classifications	Watershed
El Dovio							
El Ciprés	12 ha	N 4° 31´/W 76° 10´	1,850	1,500	SHF-M	Inceptisols/ Andisols, typic dystropepts[2] /Dystric cambisols[3]	Garrapat San Juan (Pacific)
Dagua							
V. Victoria	2 ha	N 3° 36´/W 76° 37´	1,350	1,500	Shf-pm	Inceptisols, typic dystropeps[2]/ Umbric andosols[3]	Dagua (Pacific)
El Cambio	5 ha	N 3° 36´/W 76° 37´	1,350	1,500	Shf-pm	Inceptisols, typic dystropepts[2]/ Umbric andosols[3]	Dagua (Pacific)
Paloalto	-	N 3° 35´/W 76° 37´	1,630	1,500	Shf-pm	Inceptisols, typic dystropepts[2]/ Umbric andosols[3]	Dagua (Pacific)
Queremal	-	N 3° 33´/W 76° 42´	1,390	1,500	Shf-pm	Inceptisols, typic dystropepts[2]/ Umbric andosols[3]	Dagua (Pacific)

[1]SHF-M: sub-humid forest, mountainous; SHF-PM: sub-humid forest, pre-mountainous.
[2]USDA (1999) classification.
[3]FAO, UNESCO and ISRIC (1988) classification.

Table 5. Experimental farms in Amazonia on flat and sloping topography.

Farm name	Area (ha)	Latitude and longitude	Altitude (m.a.s.l.)	Rainfall (mm/year)	Soil type classifications	Watershed
Flat topography (0-5%)						
Guajira	250	N 1° 27´/W 75° 38´	245	4,000	Ultisols, Typic Kandiudults and Typic Paleudults[1]/Haplic Acrisols[2]	Bodoquero-Orteguaza-Amazonas
Santo Domingo	60	N 1° 35´/W 75° 38	302	400	Ultisols, Oxic Dystropepts[1]/Haplic Acrisols[2]	Bodoquero-Orteguaza-Amazonas
Cartagena del Chairá	200	N 1° 45´/W 75° 42	260	4,000	Ultisols, Typic Kandiudults and Typic Paleudults[1]/Haplic Acrisols[2]	Guayas-Amazonas
Sloping topography (7-15%)						
Pekín	800	N 1° 29´/W 75° 26	265	4,500	Oxisols and Ultisols, Typic Paleudults and Typic Hapludults[1]/Haplic Ferrasols[2]	Las Margaritas-Orteguaza-Amazonas
Balcanes	60	N 1° 25´/W 75° 31	259	4,500	Ultisols, Typic Paleudults and Typic Hapludults[1]/Haplic Ferrasols[2]	Varias Quebradas-Orteguaza-Amazonas

[1]USDA (1999) classification.
[2]FAO, UNESCO and ISRIC (1988) classification.

Table 6. Experimental farms in Costa Rica.

Farm name	Area (ha)	Latitude and longitude	Altitude (m.a.s.l.)	Rainfall (mm / year)	Soil type classifications	Watershed
Esparza, Pacific Coast						
Sergio Miranda	16 ha	N 10° 02´/W 84° 42´	200	2,040	Inceptisols and Entisols[1]/Cambisols[2]	Barranca
Antonio López	68.5 ha	N 10° 02´/W 84° 42´	200	2,040	Inceptisols and Entisols[1]/Cambisols[2]	Barranca
Bernardo López	70 ha	N 10° 02´/W 84° 42´	270	2,200	Inceptisols and Entisols[1]/Cambisols[2]	San Jerónimo
Arsemio Lobo	42.3 ha	N 10° 02´/W 84° 42´	210	2,040	Inceptisols and Entisols[1]/Cambisols[2]	Destierro
Pocora, Atlantic Coast						
EARTH[3]	7.0 ha	N 10° 10´/W 83° 10´	200	3,500	Inceptisols[1]/Dystric Cambisols[2]	Destierro
EARTH	25 ha	N 10° 10´/W 83° 10´	220	3,500	Inceptisols[1]/Dystric Cambisols[2]	Destierro
EARTH	7.0 ha	N 10° 10´/W 83° 10´	350	3,500	Inceptisols[1]/Dystric Cambisols[2]	Destierro
EARTH	35 ha	N 10° 10´/W 83° 10´	400	3,500	Inceptisols[1]/Dystric Cambisols[2]	Destierro

[1]USDA (1999) classification.
[2]FAO, UNESCO and ISRIC (1988) classification.
[3]Escuela de Agricultura de la Región Tropical Húmeda, Costa Rica.

Carbon sequestration in tropical grassland ecosytems

Study sites in Costa Rica were in two contrasting ecological regions. One was Esparza, in the sub-humid seasonally dry hillsides along the Pacific Coast, and the other Pocora, close to Guapiles, in the humid lowlands along the North Atlantic Coast. These regions differ mainly in the quantity and distribution of rainfall during the year (Table 6).

In the Esparza region, the main production systems are dual-purpose and beef fattening systems based on grazing of *H. rufa* and brachiaria pastures. The research was carried out on private farms ranging in size from 16 to 70 ha. Existing cattle production systems in the Pocora region are suckling cow and calf and dual-purpose systems based on pastures as the main source of feed. The main grass species are *Ischaemum ciliare and Brachiaria* spp. The research was carried out on paddocks of the Escuela de Agricultura para la Region Tropical Humeda (EARTH) ranging from 7 to 35 ha.

Sampling design, variables measured and statistical analysis

Sampling for soil C and biomass measurements were carried out twice on each ecosytem (years 1 and 3 for the Andean Hillsides and Costa Rica and years 2 and 4 for Amazonia). The two soil C measurements were used as spacial replicates and not as time-dependent measurements. It was assumed that an additional 2 years in systems that had been established for more than 10 years would not cause measurable differences in soil C stocks.

For each land use system/site/ecosystem, linear transects of 100-300 m in Andean Hillsides and Costa Rica farms and 1000 m in the humid tropical forest of Colombian Amazonia were marked in the direction of the predominant slope gradient. Four main sampling points – major profile pits, of 150x100x100 cm, were located in the direction of the gradient, and two secondary sampling points of 40x40x100 cm, were located one on each side of the main sampling point, perpendicular to the gradient, one at maximum distance from the main sampling point within the land use area and the second one half way between the main and secondary sampling points, to allow future geo-statistical analysis of soil data. Soil samples were taken at four soil depths at each sampling point: 0-10, 10-20, 20-40, and 40-100 cm. Composite soil samples were taken at each sampling point and depth: for main sampling points by mixing 3 soil samples from three pit faces; for auxiliary sampling points by mixing 3 soil samples taken one from the given point and two from neighbouring points located within the same gradient conditions. The distances between main sampling points and from

main sampling point to secondary sampling points, slope at each main sampling point (in %) and slope shape (flat, constant, concave, convex) were recorded. Soil samples from the Colombian sites were analysed at CIAT's and those from Costa Rica at CATIE's Soils Laboratory. Analytical methods were the same for both laboratories using the following methods:
- oxidisable C: Walkley and Black (USDA, 1996: 6A1);
- total C: dry combustion at 120 °C;
- stable C: the difference between total and oxidisable C.

Additional soil determinations for all samples included: BD, pH, total N, P, CEC and soil texture (% of clay, sand and silt). Laboratory results were screened for inconsistencies as reported on Chapter 4.

For statistical comparison of land use systems in terms of soil C stocks, these were estimated on fixed soil mass according to Ellert *et al.* (2002), which adjusts soil C stocks to a constant soil weight per sampling point for a given soil depth, without subdivision in soil horizons as modified by Buurman *et al.* (2004). Following Buurman *et al.* (2004), the lowest soil mass per sampling point to 1 m depth was used as reference for each research site. Principal Component and Cluster Analyses were applied to identify relationships between soil C stocks and soil parameters and to group sampling points that were similar in soil conditions and level of soil C stocks.

C stocks in above and below ground biomass were calculated by multiplying biomass dry matter with 0.46 for C in pasture biomass and 0.48 for C in root and woody biomass, according to CATIE and Guelph (2000). Allometric equations were used to estimate aerial biomass of tree species in specific silvopastoral systems of the sub-humid and humid Tropical Forest ecosystems, Costa Rica.

For tree inventory and density, two circular plots located at each side of the main sampling point were used, with a total of 8 circular plots per system/site. Plot size for silvopastoral systems was 1000 m^2 (18 m radius). Plot size for native forest in Andean Hillsides was 100 m^2 (6 m radius) and Amazonia 1000 m^2 (18 m radius). Trees were classified by diameter-size and tree height.

Soil C changes in newly established systems on degraded land

In the Andean Hillsides one experiment was established in October 2002 at El Dovio and one in November 2002 at Dagua, with 4 treatments and 2 controls/ block, under cutting. Plot size was 100 m². The improved grass treatment was *Brachiaria* hybrid 'Mulato' (CIAT 36061) and the grass+legume treatment was *Brachiaria* hybrid 'Mulato' (CIAT 36061) + *A. pintoi*. The forage bank of consisted of *Trichantera gigantean, Morus* spp., *Erithrina edulis, Boehmeria nivea, Tithonia diversifolia*. The following fertilisers were applied at planting (kg/ha): Dagua P= 39; Ca= 185; Mg=2.5; S=3.3; N= 15; K = 12.5 =; Dovio P= 30.5; Ca= 90; Mg=2.5; S=3.3; N= 15; K = 12.5. In addition, the forage bank received 13.5 t/ha of worm compost.

In the Amazonian humid tropical forest one experiment was established in October 2002 on flat topography and one in September 2002 on sloping topography, at the 'Santo Domingo' and 'Balcanes' farms, respectively, with 4 blocks and 4 treatments per block and the degraded pasture as control. Plot size was 200 m². The area was fertilised with 2500 kg/ha of 'cal dolomitica' (25% Ca, 10% Mg) and 350 kg/ha of 'fosforita Huila' (8.7% P, 30% Ca).

The treatments were an improved grass treatment of *Brachiaria* hybrid 'Mulato' (CIAT 4624) and the grass+legume treatment was *Brachiaria* hybrid 'Mulato' (CIAT 4624) + *A. pintoi*. The forage bank colnsisted of *Gliricidia sepium, Clitoria farchildiana, Cratylia argentea, Trichantera gigantea* and *Erithryna poepigiana*.

In Costa Rica one experiment was established in September 2003 at the Esparza site, with 4 treatments per block and the degraded pasture as control. The total experimental area was 0.5 ha. The improved grass treatment was *B. brizantha* and the improved grass+legume treatment was *B. brizantha + A. pintoi*. The forage bank consisted of *Cratylia argentea*. Fertiliser was not applied.

Sampling strategy, variables measured and statistical analysis

Soil C measurements were carried out following the same methodology as described for long-established systems. The initial C stocks were estimated in the degraded pasture area before the experiment was established. The final C stock measurements were carried out at the end of 2005.

Data checking and estimation of soil C stocks were carried out following the same methodology as for long-established systems. Soil C stocks before and after 3 years of establishment of the improved systems were compared using Dunnet's test, under ANOVA models consistent with the experimental design used.

C stocks in the tree component of silvopastoral systems, forage banks and forest

The sampling plots for tree inventory and the estimation of tree biomass to calculate C were nested around the main soil profiles (3 to 4 depending on plot size) used for sampling soil C. In the center of each soil profile a plot of 25×10 m was located, in such a manner that the longer side was parallel to the longitudinal transect across the plot. Each plot of 25×10 m was divided into sub-plots for the tree inventory at different growth stages. In the forest systems tree inventories were carried out to establish the density of trees and to collect data on growth parameters (height (H) and diameter at breast height (DBH)) to estimate biomass and C stocks. Saplings (0.1 m to 0.3 m H) were counted in a 5×5 m plot, juvenile plants (> 0.3 m H, DBH ≤10 cm) in a 10×10m plot, and adults (DBH > 10 cm) in the 25×10 m plot. In the case of forest plantations (e.g. Teak – *Tectona grandis*), the size of the plot was 25×25 m. Individual trees were identified according to species and measurements of total H and DBH were taken on all adult trees (> 10 cm DBH) (Louman, 2002).

To estimate C stocks in tree resources in this project, allometric equations were developed for different systems in similar environments of the project area. More information on the development of allometric equations and estimation of C stocks in tree component can be consulted in MacDicken (1997), Segura and Kanninen (2002), Segura *et al.*, 2006) and Márquez *et al.* (2000).

To construct allometric equations for estimating biomass and hence C sequestration, destructive sampling was required to develop allometric equations to estimate biomass. This involved harvesting individual trees of different diameter classes. Each tree was divided into foliage, branches, stem (up to the first branch) and trunk (Figure 1). The foliage, large (> 25 cm diameter) and small branches and stem were weighed separately and one sub-sample from each component per tree was taken and oven-dried (60 °C, 72 hour) to obtain dry matter (DM) content. For the trunks and large branches, trunk height and large branches (>25 cm), H and diameter were measured and their volume calculated using the Huber equation $V = d^2 \times \pi/4\,h$ (Loetsch *et al.*, 1973). Samples of trunk

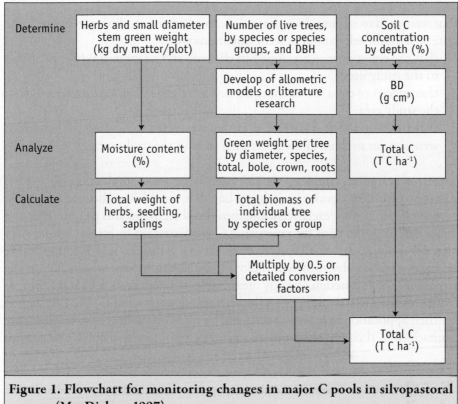

Figure 1. Flowchart for monitoring changes in major C pools in silvopastoral systems (MacDicken, 1997).

and large branches were collected and sent to the laboratory to estimate specific gravity. The equation used to estimate total biomass (TB) was:

$$TB = Vf \times SG \times BEF$$

where:

TB: total above ground biomass of tree (t DM/tree);
Vf: volume of trunk (m³/tree);
SG: specific gravity or basic density of timber (t/m³);
BEF: biomass expansion factor.

To estimate above ground C stocks of trees, the data on tree inventory and allometric equations developed in other studies were used; some of the equations that were used to estimate biomass in different systems are shown in Table 7. The criteria used for selecting allometric equations were:

1. that the equation was developed in climatic and edaphic conditions similar to the study sites;
2. that species of trees for which equations were generated were present within the study area;
3. that the ranges of tree DBH and H that were used to develop the equations were similar to those of the individual trees in the study area.

Ideally destructive sampling should have been used to develop allometric equations for each species or cohort, but this would have been a very time consuming and costly exercise, not within the objective of this project. Biomass equations form the basis for estimating C sequestration in forest, agroforestry and silvopastoral systems (Eamus *et al.*, 2000; Albrecht and Kandji, 2003). Biomass is frequently estimated employing allometric models which express the tree biomass as a function of easily measurable variables such as H and DBH, or basal area (BA) (Parresol, 1999). Allometric models have mainly been developed for their application in natural forests and forest plantations. However, there are limitations in their use in silvopastoral and agroforestry systems due to natural

Table 7. Examples of allometric equations used to estimate above ground biomass of trees in different land use systems.

Land use-species	Equation	Description	Source
Tropical forest, 900-1500 mm rainfall	$\ln Y = -2.00 + 2.32 \times \ln(dbh)$	$\ln Y$ = Logarithm-total biomass (Kg DM). dbh = Diameter breast height (cm).	Brown 1997
Tropical forest, 1500-4000 mm rainfall	$\ln Y = -2.13 + 2.53 \times \ln(dbh)$	$\ln Y$ = Logarithm-total biomass (Kg DM). dbh = Diameter breast height (cm).	Brown 1997
Tropical forest, > 2000 mm rainfall	$\ln Y = -1.97 + 2.25 \times \ln(dbh)$	$\log_{10} Y$ = Logarithm base 10 of total biomass (t/ha DM). dbh = Diameter breast height (cm).	Overman *et al.* 1994

differences in tree shape and due to alterations in their shape and architecture caused by management of trees (Andrade and Ibrahim, 2003).

Statistical analysis

An analysis of variance was conducted to determine the effect of seasons on biomass and C yield of forage banks. In the case of forest plantations and secondary forest, regression models were fitted to describe relationships between growth parameters and biomass and C yield of trees. These data were used to calculate the amount of C stored per ha.

Chapter 3. C stocks and sequestration

M.C. Amézquita, E. Amézquita, F. Casasola, B.L. Ramirez, H. Giraldo, M.E. Gómez, T. Llanderal, J. Velázquez and M.A. Ibrahim

This chapter is the core of the book. It presents the results of both the long-established systems and the short-term experiments. The presentation of results from the different ecosystems and landuse systems is not identical for each system, because the work was done by different groups of scientists, but the methodology was the same. In all cases, the reported C stocks are in t/ha/1m-equivalent, so that land uses that change the density of the soil can still be compared.

First, the results of the long-established systems will be discussed, followed by those of the short-term experiments.

C stocks in long-established land use systems

C stocks in soil and vegetation of forest and long-established pasture and silvopastoral systems (14-16 years under commercial production), as described in Chapter 2, were estimated in 2002 and 2004 for all sub-ecosystems. The estimates of 2002 and 2004 were considered as replications, because a time-effect was expected to be very small. The results do not contradict this assumption.

Andean Hillsides, Colombia

C stocks were estimated on 6 small cattle farms of 2-12 ha with poor acid soils (pH 5.2-6.2) at El Dovio (1850 m.a.s.l.) and Dagua (1400 m.a.s.l.). The results are given in Table 1 for El Dovio and in Table 2 for Dagua.

In both El Dovio an Dagua, the native forest had the largest amount of aerial biomass (data for the forest were taken from the literature), but in all systems, the soil stored larger amounts of C than the biomass. In El Dovio, the native forest also had the largest C stock in the soil, while the stocks under degraded pasture, forage bank, and *B. decumbens* pasture were not significantly different. In Dagua (Table 2), the treatments 'natural regeneration of degraded pasture' and '*B. decumbens* pasture' contained significantly larger C stocks than the degraded pasture and the forage bank.

Table 1. Stocks of total C in soil (t/ha/1m-equivalent) and biomass (t/ha) at El Dovio (1850 m.a.s.l.).

Land use system	C in soil	%	C in pasture biomass	%	C in fine roots	%	C in aerial biomass and thick roots	%	C in the system
Native forest	231[a]	75	-	-	4.6	2.0	70	23	306
B. decumbens	147[b]	95	0.9	1.0	6.3	4.0	-	-	154
Forage bank[1]	131[b]	86	-	-	4.3	3.0	17	11	152
Degraded pasture	136[b]	97	0.5	0.3	3.9	2.7	-	-	140

a-b Means followed by a different letter are statistically different at p<0.10.
[1]Trichantera gigantean, Morus spp., Erithrina edulis, Boehmeria nivea, Tithonia diversifolia.

Table 2. Stocks of total C in soil (t/ha/1m equivalent) and biomass (t/ha) at Dagua (1400 m.a.s.l.).

Land use system	C in soil	%	C in pasture biomass	%	C in fine roots	%	C in aerial biomass and thick roots	%	C in the system
Native forest	186[a]	72	-	-	2.6	1.0	70	27	259
Natural forest regeneration of degraded pasture	142[b]	97	0.8	0.8	3.2	2.2	-	-	146
B. decumbens pasture	136[b]	94	0.8	0.5	8.3	5.5	-	-	145
Forage bank[1]	90[c]	81	-	-	3.0	3.0	18	16	111
Degraded pasture	97[c]	98	-	-	2.0	2.0	-	-	99

a-c Means followed by a different letter are statistically different at p<0.10.
[1]Trichantera gigantean, Morus spp., Erithrina fusca, Tithonia diversifolia.

Carbon sequestration in tropical grassland ecosytems

Amazonia, Colombia

C stocks in soil and vegetation were estimated on farms ranging from 45 to 1600 ha in landuse systems in an area of flat and one of mildly sloping topography (Table 3). Soils in both topographies were acid with low P and N status. Although all treatments were originally present in both areas, two on sloping topography were removed because of inhomogeneous soils (see Chapter 4).

The C stocks of land uses on both topographies are given in Tables 4 and 5. As before, native forest had the highest aerial biomass, and consequently the highest total C stocks on both topographies. In all cases, the total C stock in the soil was larger than that in the aerial biomass.

On flat topography, the soils under native forest and *B. decumbens* pasture contained the smallest C stocks (Table 4). Soils under *B. humidicola* had statistically the largest C stocks, while the other land uses occupied an intermediate position. The low C stock under native forest may be due to the fact that soil under the sampled forest had worse drainage than under the other land uses, so that root penetration may have been hampered.

On sloping topography, the native forest had the largest soil C stocks, followed by *B. decumbens* + legume pasture, *B. humidicola* pasture, and natural regeneration, in that order. All differences were statistically significant. As a whole, C stocks

Table 3. Long-established land use systems in Amazonia.	
Flat topography	Sloping topography
Slopes 0-5%	Slopes 7-15%
Native forest	Native forest
B. humidicola *pasture*	B. humidicola pasture
B. humidicola + legume pasture	removed[1]
B. decumbens pasture	removed[1]
B. decumbens + legume pasture	B. decumbens + legume pasture
natural regeneration of degraded pasture	natural regeneration of degraded pasture
[1]Removed for lack of soil homogeneity (see Chapter 4).	

Table 4. Stocks of total C in soil (t/ha/1m-equivalent) and biomass (t/ha) in long-established systems, flat topography, Amazonia, Colombia.

Land use system	C in soil	%	C in pasture biomass	%	C in fine roots	%	C in forest aerial biomass	%	Total C in system
Native forest	107[a]	58.53	-	-			75.9	41.5	182.9
B. humidicola pasture	144[c]	94.7	2.5	1.6	5.6	3.7	-	-	152.1
B. humidicola + legume pasture	138[bc]	93.8	2.8	1.9	6.3	4.3	-	-	147.1
B. decumbens pasture	124[a]	97.3	1.4	1.1	2.1	1.6	-	-	127.5
B. decumbens + legume pasture	128[b]	96.0	1.6	1.2	3.7	2.8	-	-	133.3
Natural regeneration of degraded pasture	134[bc]	96.8	1.7	1.2	2.7	2.0	-	-	138.4

[a-c]Means followed by a different letter are statistically different at p<0.10.

Table 5. Stocks of total C in soil (t/ha/1m-equivalent) and biomass (t/ha) in long-established systems, mildly sloping topography, Amazonia, Colombia.

Land use system	C in soil	%	C in pasture biomass	%	C in fine roots	%	C in thick roots, trunks and leaves	%	Total C in system
Native forest	181[a]	58.5	-	-	-	-	128.5	41.5	309.5
B. humidicola pasture	159[c]	96.1	1.4	0.8	5.1	3.1	-	-	165.5
B. decumbens + legume pasture	172[b]	97.8	1.2	0.7	2.7	1.5	-	-	175.9
Natural regeneration of degraded pasture	129[d]	96.8	1.2	0.9	3.0	2.3	-	-	133.2

[a-d]Means followed by a different letter are statistically different at p<0.10.

Carbon sequestration in tropical grassland ecosytems

on sloping topography were higher than on flat topography. This may also be a result of slightly impeded drainage, and seasonal water logging of the flat parts.

Factors affecting soil C stocks in long-established Amazonian ecosystems

The previous tables suggest that land use systems are the main cause of differences in C stocks. However, because some plots are found on strongly sloping land, and repetitions are sometimes at considerable distance, these factors may also play a significant role. Factor analysis allows the recognition of such influences (Chapter 5, the effects of soil properties on C stocks, also analysed by factor analysis, will be discussed). Table 6 summarises the main factors that influenced C stocks in the Colombian ecosystems.

Table 6 shows that, apart for the Amazonia flat topography, where land use resulted in only two significant groups of soil C stocks, land use was the main factor influencing total soil C stocks. In the Andean area of El Dovio and Dagua, slope position played a significant role, and in both Amazonia areas, there was a distinct repetition effect, indicating that replications, if at considerable distance, may react differently. In Amazonia flat topography, land use explains 51% of the

Table 6. Variance explained by factors affecting total soil C stocks (t/ha/1m-equivalent) in long-established systems of Colombia.

Factors controlled by the sampling design	Explained variance (%)[1,2]			
	El Dovio	Dagua	Amazonia flat	Amazonia sloping
Land use system	60***	60***	43***	74***
Repetition (land use system)	3		10**	7***
Position (Rep x landuse)	2			7
Slope	11**	11***		
Non-explained variation	24	29	47	12

[1]Probability of significance of F-test: *** p<0.01; ** 0.01<p<0.05; * 0.05<p<0.10.
[2]Estimated from the analysis of variance as (Factor sum of squares/total sum of squares) x 100.

variation in stocks of stable C (not shown). Unexplained variation is the second largest factor in all systems. Unexplained variation can be due to unstudied factors such as local variation in vegetation, but also to random spatial variation. The latter factor, which can be studied by geostatistical methods, has not been included in the present research.

On both topographies in Amazonia, 60% of mean soil C was found at a depth of 0-40 cm and 40% at 41-100 cm, which is comparable with the finding of Fisher *et al.* (1994), who measured C storage of 237 t/ha under a 6-year-old *Andropogon gayanus-Stylosanthes capitata* pasture, with about half of it in the 41-100 cm deep soil layer in the llanos of Colombia.

Costa Rica

The two experimental areas in Costa Rica were Esparza, (Pacific coast, 200 m.a.s.l.; 2,000 mm precipitation/yr) and Pocora (Atlantic coast, 200-400 m.a.s.l.; 3,500 mm precipitation/yr). The land uses evaluated in both areas are listed in Table 7.

In Esparza, there are two groups of land uses with respect to soil C stocks. Soils under teak (*Tectona grandis*) plantation, secondary forest, *H. rufa* and degraded weedy pasture had significantly higher amounts of soil C than those under native forest, forage bank, silvopastoral system, *B. decumbens* pasture and degraded weedy pasture. Total soil C under *H. rufa* pasture was almost double that under *B. decumbens* monoculture and under the silvopastoral system. The differences in total C between *H. rufa* and brachiaria pastures is probably related to *H. rufa* pastures being older and that these pastures were traditionally burned annually, which is known to increase the amount of stable C stocks. Although fire releases considerable amounts of C to the atmosphere, it also converts C in woody material into charcoal and C in non-lignified parts into char, which increases the long term C storage in soils (Minami *et al.*, 1993).

In Pocora, *I. cilare* pasture, improved pastures (*B. brizantha* monoculture and in association with *A. pintoi*) and *A. mangium* + *A. pintoi* had significantly (49-140%) higher total C stocks than the degraded pasture. Native forest had relatively low total soil C compared to *I. cilare* pasture (Table 7).

In the native forest of Esparza, the amount of C sequestered in tree biomass was about equal to the amount of C stored in the soil. In the secondary forest,

Table 7. Total C stocks in soil (t/ha/1m-equivalent) and in tree biomass (t/ha) in Esparza and Pocora.

Land use	Min. age of use	C in soil	C in tree aerial biomass	Total C
Esparza				
Degraded weedy pasture	>50	154.8[a]	-	154.8
H. rufa	>100	220.9[b]	-	220.9
B. decumbens	8	109.6[a]	-	109. 6
Silvopastoral system[1]	9	120.0[a]	17.2	137.15
Forage bank[2] (13)	13	124.3[a]	-	127.56
Teak plantation	9	222. 8[b]	92.4	315.18
Secondary forest	11	226.0[b]	58.3	284.29
Native forest		94.4[a]	99.9	194.34
Pocora				
Degraded pasture	>70	107.9[d]		107.9
I. ciliare	>70	254.4[f]		254.4
B. brizantha	>19	153.0[e]		153.0
Silvopastoral system[3]	15	160. 9[e]	12.8	173.5
B. brizantha + A. pintoi	16	186.8[e]		186.8
Native forest		141.3[de]	174.2	315.5

[a-f]Means followed by a different letter are statistically different at $p < 0.10$.
[1]*B. decumbens* with trees.
[2]*Leucaena leucocephala* and *Cratylia argentea*.
[3]*Acacia mangium* and *A. pintoi*.

teak plantations and silvopastoral system, the amount of C sequestered in tree biomass represented 25.8, 41.5, and 14.3%, respectively, of the amount of total C stored in the soil (Table 7).

In Pocora, the native forest had a higher amount of C sequestered in biomass of trees than in the soil (Table 7). The amount of C sequestered in the biomass of *A. mangium* in the silvopastoral system represented 8% of the amount stored in the soil. *A. mangium* is a fast growing tree; in the silvopastoral system it had an annual rate of C fixation of 1.6 t/ha/yr in the tree biomass.

Total C in the aerial biomass of the various pastures at Esparza and Pocora were measured during three consecutive seasons. The results are reported in Table 8. The annual production appears to be higher under the more humid conditions of the Atlantic zone. It should be kept in mind, though, that in grasslands the production of subsurface litter (root litter) is more important to C stocks than surface DM.

In contrast to Chapter 5 of this book, the Principal Component analysis for the Costa Rican sites was carried out for properties quantified to 1 m depth equivalent. Only Factors (or Principal Components-PC) with Eigenvalues greater than 1 are shown in Table 9, meaning that their contribution to explain variation is greater than that of the original variables. Three Principal Components explained 91% in Esparza and two explained 69% in Pocora of the variation in the original variables. In PC1 and PC2 for Esparza and PC 1 for Pocora, there is a strong relationship between total C, oxidisable C, and total N. In the F1F2 diagrams of Esparza and Pocora and in the F1F3 diagram of Esparza, oxidisable C plots between Total C and clay, suggesting that clay influences the amount of oxidisable C. In the F1F2 diagram of Esparza, CEC plots close to clay and is not clearly related to Total C, while for Pocora, CEC plots close to sand, which seems illogical. In the F1F3 diagram of Esparza, CEC plots between *sand* and *Total C + clay*, which appears more logical. Unknown variations in clay mineralogy and SOM humification may have caused the deviations of the Pocora data.

Table 8. Mean quantity per year of C (t/ha) in available DM of pastures in 2003-2005 in Esparza and Pocora.

Esparza		Pocora	
Landuse	Mean	Landuse	Mean
		A. mangium + A.pintoi	0.24 ± 0.07[d]
B. decumbens pasture	0.86 ± 0.21[a]	B. brizantha + A. pintoi	1.3 ± 0.19[a]
H. rufa pasture	0.76 ± 0.15[a]	B. brizantha	1.21 ± 0.11[a]
Weedy pasture	0.56 ± 0.16[b]	I. ciliare	0.99 ± 0.08[b]
Silvopastoral system	0.77 ± 0.27[a]	Degraded pasture	0.65 ± 0.07[c]

[a-d]Means in the same column followed a different letter are statistically different at p<0.05.

Table 9. Principal component analysis of data from Esparza and Pocora. Association between soil variables and soil C at 0-100 cm depth.

	Esparza			Pocora	
Variable	PC1 (46%)	PC2 (25%)	PC3 (20%)	PC1 (48%)	PC2 (21%)
Sand (mg/ha)	-0.33	0.63	0.10	0.30	-0.40
Clay (mg/ha)	0.33	-0.46	-0.51	0.11	0.72
CEC	0.13	-0.33	0.78	-0.24	0.44
Total C (mg/ha)	0.54	0.28	0.10	0.56	0.16
Oxidisable C (mg/ha)	0.42	0.44	-0.22	0.53	-0.16
Total N (mg/ha)	0.54	0.12	0.24	0.49	0.28

Soil C sequestration by newly established improved land use systems on degraded land

Replicated small plot experiments were established on degraded pasture land to monitor the changes in soil C stocks during the experimental period of 3.4 years. The treatments consisted of improved grassland with different grass monocultures, grass plus legume mixtures and forage banks in comparison with the original and the naturally regenerating degraded pasture. The vegetation cover of the original degraded pastures consisted of combinations of grasses, herbs and shrubs.

Initial C stocks were estimated in the degraded pasture before the establishment of the experiments and final C stocks were estimated at the end of the experimental period for each improved option. C stock changes were calculated based on fixed-mass profiles of 1m depth-equivalent.

Andean Hillsides, Colombia

The improved systems for the Andean hill sides were:
1. *Brachiaria* hybrid 'Mulato' monoculture;
2. *Brachiaria* 'Mulato' + *A. pintoi*;
3. Forage bank regularly cut for animal feed consisting of *Erythrina fusca*, *Trichantera gigantea* and *Tithonia diversifolia*.

The results for Dagua and El Dovio are reported in Tables 10 and 11.

In Dagua both *Brachiaria* hybrid Mulato and 'natural regeneration of degraded pasture' showed statistically significant increases in soil C stocks (Table 10). The forage bank did not show a significant increase, probably due to the fact that the forage bank species contributed only small amounts of surface and subsurface litter to the soil.

Table 10. Soil C stocks and increments (t/ha/1m-equivalent) in degraded pasture and improved land use systems at Dagua. Means of 4 observations.

System	Soil C stock	In 3.4 years	Per year
Brachiaria hybrid Mulato	187.1	38.2***	11.2
Brachiaria 'Mulato' + *A. pintoi*	172.1	23.2 ns	6.8
Forage bank	154.5	5.6 ns	1.6
Natural regeneration of degraded pasture	176.5	27.6***	8.1
Degraded pasture (reference)	148.9	-	-

*** Statistically significant with $p < 0.001$; ns statistically non-significant.

Table 11. Soil C stocks and increments in soil C stocks (t/ha/1m-equivalent) in degraded pasture and improved landuse systems at El Dovio. Means of 4 observations.

System	Soil C stock	In 3.4 years	Per year
Brachiaria hybrid Mulato	153.6	-11.2 ns	-3.3
Brachiaria 'Mulato'+ *A. pintoi*	188.5	23.7*	7.0
Forage bank	166.3	1.5 ns	0.4
Natural regeneration of degraded pasture	153.3	-11.5 ns	-3.4
Degraded pasture (reference)	164.8	-	-

* Significant with $p < 0.05$; ns statistically non-significant.

Although not statistically significant, some of the treatments at Dovio (Table 11) appear to decrease soil C stocks. The fact that also the natural regeneration of the degraded pasture falls into this category suggests that regrowth at this altitude may require a longer period of time. Only the mixture of *Brachiaria* hybrid Mulato + *A. pintoi* showed a statistically significant increase in soil C stocks.

Amazonia, Colombia

Two replicated small plot experiments, one in flat topography and one in sloping topography, were established on degraded pasture areas to monitor the changes in soil C accumulation of newly established improved systems during 3.3 years. The treatments were:

1. *Brachiaria* hybrid CIAT 4624 in monoculture;
2. *Brachiaria* hybrid CIAT 4624 + *A. pintoi;*
3. forage bank cut for animal feeding;
4. natural regeneration of the degraded pasture.

The existing degraded pasture was used as the reference. The results are given in Tables 12 and 13.

On flat topography (Table 11) all improved land uses resulted in a significant increase in C stocks, while the grass+legume association had significantly the highest C sequestration. There were no significant differences between the grass-

Table 12. Increments in total soil C stocks (t/ha/1m-equivalent) in improved land use systems on flat topography in Amazonia. Means of 4 observations.

System	Soil C stocks	In 3.3 years	Per year
Brachiaria pasture	147.3	11.6***	3.5
Brachiaria + *A. pintoi* pasture	159.3	23.6**	7.2
Forage bank	146.8	11.1**	3.4
Natural regeneration of degraded pasture	146.1	10.4**	3.2
Degraded Pasture (reference)	135.7		-

Statistical significance according to the Dunnet test: *** $p < 0.001$; ** $p < 0.01$; ns not significant.

Table 13. Increments in total soil C stocks (t/ha/1m-equivalent) in improved land use systems on sloping topography in Amazonia. Means of 4 observations.

System	Soil C stocks	In 3.3 years	Per year
B. hybrid CIAT 4624 pasture	160.4	25.5 ***	7.7
B. hybrid CIAT 4624 + *A. pintoi* pasture	141.0	6.1 ns	1.8
Forage bank	153.1	18.2 **	5.5
Natural regeneration of degraded pasture	165.0	30.1 ***	9.1
Degraded Pasture (reference)	134.9		-

Statistical significance according to the Dunnet test: *** p< 0.001; **p<0.01; ns not significant.

alone system, the forage bank and the natural regeneration of a degraded pasture. The fact that all treatments showed significant differences with the original degraded pasture may be due to the fact that the soils of the experimental plots were relatively homogeneous.

On sloping topography (Table 13), the '*B.* hybrid CIAT 4624 pasture', the 'natural regeneration of degraded pasture' and the 'forage bank' had statistically significant C sequestration, but that of the forage bank was significantly lower than the other two.

Costa Rica

In Costa Rica new experiments were restricted to the Esparza area in the Pacific zone. The duration of the experiment was 3 years; sampling for C stocks was carried out in 2003 and 2006. The results are given in Table 14.

The amount of soil C sequestered in Esparza was significantly higher for grass monoculture pastures (*H. rufa* and *B. brizantha*) and the grass legume mixture (*B. brizantha* + *A. pintoi*) compared to the *C. argentea* forage bank and 'natural forest regeneration', but there were no differences between the grass based-pastures, nor between the forage bank and natural forest regeneration (Table 14).

Table 14. Total soil C stocks (t/ha/1m-equivalent) in improved land use systems in Esparza at the beginning and the end of the experimental period.

Landuse	C stock in 2003	C stock in 2006	C sequestered per year
B. brizantha pasture	145.8 ± 10.5	156.3 ± 9.3	3.5[a]
B. brizantha + *A. pintoi* pasture	141.9 ± 11.1	154.2 ± 3.8	4.1[a]
H. rufa	144.0 ± 12.5	155.2 ± 15.0	3.7[a]
Forage bank of *Cratylia argentea*	131.4 ± 5.7	137.4 ± 7.6	2.0[b]
Natural forest regeneration	139.9 ± 16.7	145.9 ± 8.8	2.0[b]

[a-b]Numbers followed by different letters are significantly different at $p < 0.10$.

The relatively low C sequestered by the forage bank compared to the grass treatments may be due to the three facts: (1) the grasses are C_4 plants, which have a relatively high capacity for fixing C under good soil and climatic conditions; (2) the forage banks are cut regularly to feed the animals and produce little litter on the soil, and (3) grasses produce large amounts of litter in the soil, which contributes more directly to soil C stocks.

C stock distribution in the soil profile in the various ecosystems

Table 15 shows the mean, standard deviation and CV (%) of total C stock for the upper (0-40cm), lower (40-100cm) and the whole soil profile (0-100 cm) for each research site in all ecosystems studied. Means are based on original C stock data, expressed at fixed soil depth, from all land use systems sampled at each research site during the two C sampling cycles (2002-2004 for Andean Hillsides and Costa Rica ecosystems, and 2003-2005 for the humid tropical forest ecosystem, Amazonia, Colombia).

Data show that soil C stocks are higher at higher altitudes, as observed in the Andean Hillsides ecosystem, Colombia. Both lower altitude ecosystems – the humid tropical forest, Amazonia, Colomnbia, and the humid and sub-humid tropical forest, Costa Rica – show similarly lower stocks. On both Andean Hillsides sites, and on both Costa Rican ecosystems, more than 60% of mean soil C was found at the upper soil layer. However, on both topographies in

Table 15. Total C stock distribution in the soil profile for all ecosystems.

Soil depth (cm)		Andean Hillsides, Colombia		Humid tropical forest, Amazonia, Colombia		Humid and sub-humid tropical forest, Costa Rica	
		El Dovio[1]	Dagua[2]	Flat[3]	Sloping[4]	Pocora, sub-humid[5]	Esparza, humid[6]
0-40	Mean (t/ha)	129	98	77	84	103	117
	% from whole profile	64	61	52	56	68	78
	Sd (t/ha)	26	25	8	9	26	40
	CV (%)	20	26	11	11	25	34
40-100	Mean (t/ha)	74	62	72	67	49	33
	% from whole profile	36	39	48	44	32	22
	Sd (t/ha)	24	21	9	17	19	24
	CV (%)	32	34	13	26	38	74
0-100	Mean (t/ha)	203	160	149	151	152	150
	% from whole profile	100	100	100	100	100	100
	Sd (t/ha)	43	40	15	17	43	60
	CV (%)	21	25	10	11	28	40

[1]N (No. of sampling points or soil pits) = 96 (4 land use systems x 24 soil pits/system).
[2]N = 144 (6 systems x 24 soil pits/system).
[3]N = 189 (7 systems x 27 soil pits/system).
[4]N = 162 (6 systems x 27 soil pits/system).
[5]N = 90 (6 systems x 15 soil pits/system).
[6]N = 105 (7 systems x 15 soil pits/system).

Amazonia, around 50% of mean soil C was found at a depth of 0-40 cm, which is comparable with the finding of Fisher *et al.* (1994), who measured C storage of 237 t/ha under a 6-year-old *Andropogon gayanus-Stylosanthes capitata* pasture, with about half of it in the 41-100 cm deep soil layer in the llanos of Colombia.

General discussion of C stocks and sequestration

Long-established land use systems

The following trends can be observed:
- In pasture and silvopastoral systems, the soil C stock represented more than 80% of the total C in the ecosystem. Therefore, even small increases of the soil-C stock in such systems contribute significantly to C sequestration.
- In the Andean Hillsides, Colombia, the effect of improved land uses systems on soil C stocks is more marked at low than at high altitudes, although the level of stocks may be higher at high altitudes. Natural regeneration and *B. decumbens* pastures appear to sequester C most efficiently.
- In Colombian Amazonia, soil C stocks were generally higher on sloping than on flat land. The effect of each improved land use system is not equal on sloping and flat land. On sloping land, *B. humidicola* performed better than *B. decumbens*, while the reverse was true on flat land. The admixture of a legume may have either a positive or a negative effect on C stocks
- In all experiments in Colombia, land use was the main factor explaining changes in C stocks.
- In the humid tropical region of Costa Rica, improved pastures and silvopastoral systems had higher soil C sequestration than degraded pasture and native forest, indicating that well managed pastures have an important role in mitigating green house gas emissions.
- In the sub-humid zone of Costa Rica, pastures and silvopastoral system had similar or higher amounts of C sequestered in the soil compared to native forest
- The improved grasses and legume mixtures have a relative large percentage of C sequestered in the fine root biomass which is an important source of C cycling in the soil system.
- In Costa Rica, improved gras monocultures and grass-legume mixtures fixed relatively larger amounts of C in the soil compared to forage banks with shrubs and trees
- In Costa Rica, total C was negatively correlated with pH, and positively with clay content.

Short-term experiments
- The short-term experiments are far from reaching equilibrium, and the accumulation rates calculated for C stocks can not be extrapolated far into

the future. On the other hand, some systems may show increased rates of C accumulation beyond the experimental period.

- The results of the short-term experiments partially reflect – and therefore support – those of the long-established land use systems.
- In the Andean Hillsides, the land use systems *Brachiaria* hybrid mulato, *Brachiaria* mulato + *A. pintoi*, and natural regeneration of degraded pasture were most effective in sequestering C in the soil. Significant effects were more common at low altitude, possibly due to greater soil heterogeneity and slower growth at higher altitude.
- In Amazonia, virtually all improved land use systems caused a significant increase in soil C stocks within the experimental period of 3.4 years, with yearly increments varying between 3 and 9 tons of C per hectare.
- In Costa Rica, all tested land use systems showed a significant increase in C stocks within the experimental period of 3 years. Increments varied between 2 and 4 tons of C per hectare.

Soil C measured in the native forest was relatively low in the humid tropical ecosystem in Costa Rica and this is consistent with what has been reported in the literature. Studies conducted in Puerto Rico by Lugo *et al.* (1986) showed that pastures contained similar or greater amounts of soil C than adjacent mature forest. In Brazil, under certain circumstances, pastures replacing forests have been accumulating more C in soils than adjacent forests (Fearnside and Barbosa, 1998; Tarré *et al.*, 2001; Desjardins *et al.*, 2004). In northeastern Costa Rica, soil C pools to 30 cm depth ranged from 26% lower to 23% higher in pastures than in paired forests (Powers and Veldkamp, 2005).

The results of this study are consistent with other studies, which showed that native forest and other forest systems sequester large amounts of C in the tree biomass compared to that sequestered in the soil system, and some authors consider this as a mechanism of storage. Storing C in trunks of trees may represent one way of increasing permanent C stocks under sustainable harvesting and processing of timber. In grass monocultures and legume mixtures a higher percentage of C is stored in the soil. The permanence of this C depends on production and management but is usually larger than that of living (tree) biomass.

Silver *et al.* (2000) reviewed the literature and showed that aboveground biomass in new forests increased at a rate of 6.2 t/ha/yr during the first 20 years, and at a rate of 2.9 t/ha/yr over the first 80 years of regrowth. C sequestration rates have been found encouraging in secondary forest fallows (5 to 9 t C/ha/yr); complex agroforests (2 to 4 t C/ha/yr); simple agroforests with one dominant species such as oil palm, rubber or *Albizia falcataria* (7 to 9 t C/ha/yr). The lower C sequestration rate of some agroforestry systems in relation to natural secondary forest is partly because farmers use some products (Pandey, 2000).

Data of this project indicate that mean soil C in native forests and in long-established pastures of all ecosystems studied were about the same (157 vs. 160 t/ha). However, mean total C (in soil and above ground biomass) in forest was 40% more than in grasslands (261 vs. 162 t/ha). Newly established pastures and natural forest regeneration sequestered on average in all ecosystems 6 t C/ha/yr in the soil during 3.4 years.

Land use changes from forest to pastures are reported to cause a loss of about 9.5 ton C/ha (Fearnside *et al.*, 1998). With changed land use a new equilibrium is established, and this equilibrium varies from one place to the other. Whether a tropical pasture is a sink or a source of C strongly depends on its management.

Grass monoculture pastures are generally managed extensively without the use of fertilisers and these pastures degrade in time, which results in significant loss of C stocks. Veldkamp (1994) found a net loss of 2-18% of C stocks in the top 50 cm of equivalent forest soil after 25 years under pasture in lowland Costa Rica. Lugo *et al.* (1986) and Veldkamp (1993) found that the conversion of forest to arable crops resulted in greater losses of soil C than conversion to pastures. However, the combination of a deep-rooted grass with a nitrogen-fixing legume can increase nutrient cycling, animal production, soil biological activity and C sequestration, especially in the surface where the C storage take place (Fisher *et al.*, 1994). The apparent rate of C accumulation (0-100 cm depth) under the grass plus legume pastures (1.2 t/ha/yr) was 70% higher than under the grass-only pastures (0.7 t/ha/yr) (Tarré *et al.*, 2001). Well-managed grass legume mixtures (e.g. *B. brizantha* + *A. pintoi*) have been shown to increase C sequestration in soil to levels comparable to or higher than those measured in soils of native forest (Ibrahim and Mannetje, 1998). In Colombian Amazonia, C in soil to 1 m depth equivalent was 144 and 159 t/ha for *B. humidicola* on flat and sloping land, respectively, and 138 t/ha for *B. humidicola* + *A. pintoi* on sloping land, a significant increase with respect to the degraded pasture, and even higher on flat land than the soil C stock under native forest. In the humid tropics, legumes can fix more than 100 kg N/ha/yr which is an important source for sustaining high productivity of the companion grass, resulting in higher C sequestration (Ibrahim and Mannetje, 1998). Well-managed pastures of introduced grasses of African origin accumulated C in the soil at rates close to 3 t/ha/yr. Even mismanaged *Andropogon gayanus* accumulated C in the soil at this same rate (Fisher and Thomas, 2004).

In grassland systems, most of the C changes occur in the top soil layer (Degryze *et al.*, 2004). When pasture is established after crops, soil C stocks increase to below 100 cm depth but the fractional increase decreases with depth (Guo and Gifford, 2002).

Integration of multipurpose trees and shrubs (e.g. *Erythrina* spp., *Gliricidia sepium*, *Acacia mangium*) in pastures established in humid and sub-humid zones resulted in significant increases in soil organic matter, N and P, which results in increased DM production (Bolivar *et al.*, 1999). Conversion of forest to uncultivated grazing land did not, on average, lead to loss of soil C.

Highly productive agroforestry systems, including silvopastoral systems, can play an important role in C sequestration in soils and in the woody biomass. Well managed silvopastoral systems can improve overall productivity (Bustamante *et al.*, 1998; Bolivar *et al.*, 1999), while sequestering C (López *et al.*,1999; Andrade, 1999), which is a potential additional economic benefit for livestock farmers. The amount of C fixed in silvopastoral systems is affected by the tree/shrub species, density and spatial distribution of trees, and shade tolerance of herbaceous species (Nyberg and Högberg, 1995; Jackson and Ash, 1998).

On the slopes of the Ecuadorean Andes, total soil C increased, with respect to open pasture, from 8% under open *Setaria sphacelata* pasture to 11% beneath the canopies of *Inga* sp. but no differences were observed under *Psidium guajava*. Soils under *Inga* contained an additional 20 t C/ha in the upper 15 cm compared to open pasture (Rhoades *et al.*, 1998). Other studies have shown that total C in silvopastoral systems varied between 68 and 204 t/ha, with most C stored in the soil, while annual C increments varied between 1.8 to 5.2 t/ha (Andrade, 1999; Mora, 2001). In the humid tropics of Costa Rica, Andrade (1999) estimated C fixation rates of silvopastoral systems with improved grass pastures (*Panicum maximum* and *B. brizantha*) and exotic trees *Acacia mangium* and *Eucalyptus deglupta*, of 3.7 to 4.7 t C/ha/yr and a large percentage (> 70%) of C fixed in the system was in the tree biomass.

Many variables affect soil C content and C sequestration, specifically above-and below-ground litter production, litter placement, litter decomposability, soil texture and structure, physical protection through intra-aggregate or organo-mineral complexes, rainfall, temperature, farming system, soil management and age of the land use after forest clearing. Changes may occur, e.g. by increasing below-ground inputs, by enhanced surface mixing either by soil organisms or by mulching (Post and Kwon, 2000; Lal, 2004). Management may therefore be a key factor influencing soil C stocks.

Improved tropical pastures and silvo-pastoral systems generally increase soil C stocks, but the quality of management of tropical pastures decides whether the soils under such land uses present a source or a sink of atmospheric C. In well managed pastures in formerly forested areas, significant amounts of litter (roots and leaf litter) are recycled in the system which results in accumulation of soil organic C (Neil *et al.*, 1997). However, according to these authors and to Tarré *et al.* (2001) in the State of Bahia of Brazil samples taken to a depth of 100 cm showed that below 40 cm depth there was no significant contribution of the brachiaria-derived C. This is contrary to findings by Fisher *et al.* (1994), and data of this project. Roscoe *et al.* (2001) also found that after 23 years of brachiaria pastures in Minas Gerais (Brazil) the grass influence on C storage reached deeper than 1 m.

If pasture growth is reduced due to lack of maintenance fertilisation and/or overgrazing, C accumulation rates decrease, eventually to the point where soil organic matter decomposition rates exceed the deposition rate and total C stocks would decline over time. Such a decline, shown by a decrease of forage production and an increase in weed invasion, is accompanied by a decrease in root biomass and by a concentration of shallow root biomass (Müller *et al.*, 2004) and will thus lead to a loss of C from deeper layers.

Fisher *et al.* (2007) published a thorough, comprehensive in-depth review of C in grazing systems in South America (cerrados in Brazil and savanna in Colombia). They concluded that litter production and quality are key components of C sequestration and that net above ground primary production of introduced African grasses is much greater than previously known. These grasses also have a below ground net primary production that is about 75% of the above ground net primary production. Degraded pastures and native vegetation are little different in C stocks, but well managed brachiaria pastures have nearly always more soil C than native vegation that they have replaced.

Chapter 4. Analysis of soil variability and data consistency

P. Buurman and O. Mosquera

Introduction

Soils in general show a large spatial variability. Such variability may obscure effects of different crops or management practices if insufficient field information is available. It is therefore essential to establish whether ecosystems that are to be compared are indeed on similar soils, and to what degree. Homogeneity of the soil is a prime requisite for extrapolation of data.

In this project spatial variation played a role at two levels. In the long-established land use systems, where the effect of long term differences in land use on C stocks was evaluated, use was made of existing fields that occurred at considerable distance from each other. The chances that such soils are different are larger than when fields are close together. The newly established land use field trials, however, consisted of 16 small adjacent plots. Usually soil variability at short distance is smaller, but under special circumstances, such as landscapes with eroded volcanic soils, short-distance variation can also be considerable.

Soils were compared in the field by studying their morphology, including properties such as structure, aggregation, and consistency. Although field studies provided a preliminary judgment of soil homogeneity, laboratory analyses were necessary to substantiate these judgments. Laboratory analyses were carried out in two categories:
1. analyses to describe the soil (pH, Cation Exchange Capacity (CEC), texture (sand-silt-clay), and fertility parameters such as available N, P and K);
2. analyses for the calculation of C stocks (Bulk Density (BD), total C, and oxidisable C).

Some soil properties are closely related and such relations can be used to ascertain consistency of analytical data. This is hardly the case in properties of the first category. Contents of elements such as Ca, K, T, and P, which are taken up and recycled by plants and leached from the soil matrix by rainwater, are extremely variable.

For properties that are used to calculate C stocks, consistency checks of analytical data are possible. Soil BD is notoriously variable and, therefore, BD measurements are always carried out in triplicate. In the present project, the replications were taken from different sides of each soil pit, which increases reliability.

If only one organic C determination were available, it would be virtually impossible to judge data quality and identify unreliable measurements. In all samples, both total C (by element analyser) and oxidisable C (by Walkley and Black (1934) wet oxidation, WB) were determined. In most soils, under the same land use, there is a very high correlation between the two methods, differences being due to some standard conversion factors in the WB method and to C that is difficult to oxidise, such as clay-bound complexes and charcoal. In a homogeneous set of soils, clay-bound stable C is roughly constant and will not cause differences between samples. Erratic differences between the two C determinations arise when some samples contain concentrations of extremely stable C, such as charcoal. Charcoal is not homogeneously distributed in soil horizons, and samples with charcoal will show a larger-than-normal difference between total and oxidisable C. In the present research, charcoal would interfere with calculations of changes in C stocks, and samples that are likely to contain charcoal have been excluded.

The methods used in this project for soil data evaluation will be analysed and a number of examples with their proper interpretation will be presented. There are two approaches to quality control of soil analyses:
1. the use of reference samples in the laboratory;
2. a check on the internal consistency of the results upon delivery.

It is good laboratory practice to use reference samples. Such samples are analysed with each batch of new samples, and if the analyses of the reference samples deviate from the acceptable norm, the analysis of the whole batch is repeated. This ensures the reproducibility of the laboratory procedures, without which large data sets are unreliable. Even when reference samples are used, some soil samples may give erratic results, sometimes because samples are accidentally switched, but largely due to unknown factors. Such erratic results are a special problem if data sets are analysed statistically and it is therefore of the utmost importance to identify and eliminate these. The only way to find incidental errors is to look at the consistency of the data for each sample or sample set. There are many logical connections between soil data, and if most samples confirm such connections while a few do not, it is likely that the latter samples have analytical errors. We will use internal consistency of data to find incidental errors and to judge the homogeneity of the set of soils used to compare different land use systems.

Homogeneity of soils

A first and simple approach to ascertain soil homogeneity is the comparison of BD profiles. Most soils have an increase in BD with depth that is related to decreasing root and biological activity. BD is influenced by cultivation practices (e.g. tillage/no tillage) but in soils with equal land use differences tend to be less than 0.3 kg.dm^{-3} in similar horizons. Larger differences can be due to differences in land use or to, for example exposure of different subsoil layers by erosion.

In most soils, the CEC largely depends on clay fraction and soil organic matter (OM). The CEC of the clay fraction is mostly independent of pH and only depends on clay mineralogy, while that of the OM increases with pH. At a fixed pH, there is a linear relation between OM content and its contribution to CEC as long as the humification degree of the OM is similar. In soil profiles, the humification in the surface horizon may be lower than in the deeper horizons, but in all but litter layers and hydromorphic soils this effect is negligible. To characterise the C-seq. samples, CEC was determined by the Na-acetate method at pH 7 (fixed pH; for all chemical methods see SSAC, 1996). Because a fixed pH eliminates the pH-effect on CEC of OM, the relation between soil CEC and the clay and organic C contents is:

$$CEC(7)_{soil} = a \times clay + b \times C_{total}$$

in which CEC(7) is the cation exchange capacity of the soil (cmol$_c$/kg) measured at pH 7. Clay and C$_{total}$ contents are expressed in g/kg, and a and b are the contributions of the organic and clay fraction, respectively, expressed as cmol$_c$/kg.

An easy graphic method to find the relation between C content and its CEC is to plot organic C content of a sample on the X axis, and the CEC divided by the clay content on the Y axis (Figure 1). In a homogeneous set of samples, the slope of the (linear) regression between sample points gives the dependence of the CEC on C content (the slope is the CEC per gram of organic C), while the cut-off on the Y axis is the CEC of the clay fraction (the CEC per gram clay when organic C equals zero). A heterogeneous set of soil samples will show a

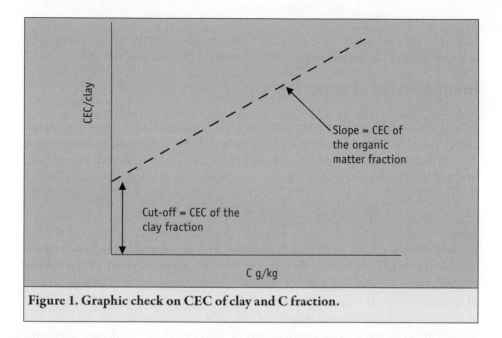

Figure 1. Graphic check on CEC of clay and C fraction.

low correlation between data points, or different correlations for samples from different fields.

These relations were tested for the Amazonian and Andean Hillsides data sets (Colombia) and for Costa Rica. In the present chapter only the data from Colombia will be referred to. At the El Dovio site (Andean Hillsides), the clay fraction probably consists of amorphous Al-silicates (allophane). This may partially invalidate the assumption that the CEC of the clay fraction is constant with depth, because (a) allophane is difficult to disperse and the measured clay fraction may therefore be underestimated, and (b) the CEC of allophane depends, in addition to pH, on its chemical composition.

Consistency of C analyses

C determined by the two methods is compared graphically for each set of laboratory results. In consistent data sets, a clear correlation will arise between the two determinations. If oxidisable exceeds total C, this is clearly an error. Outliers that do not conform to the general correlation between the two parameters, are easily spotted. Although this is not unequivocal, total C determinations tend to be more reliable than those of oxidisable C, which involve a relatively large number of steps, so that it might be sufficient to eliminate just the data for oxidisable C of outliers. In general, we have eliminated all outliers.

Results

Homogeneity of soils in trial fields

Amazonia

The Amazonia data set consists of four subsets: long existing and newly established land use systems in flat and in sloping areas. If total C is plotted against 100×CEC/clay for all Amazonian samples, the result will be a large cloud of points that does not seem to have any significance (Figure 2). When the data from flat and sloping positions are separated, it appears that the former is more homogeneous than the latter (Figures 3 and 4). Within each set of data, there are high correlations for each land use (Table 1), except for degraded pasture in sloping land, indicating less homogeneous soil.

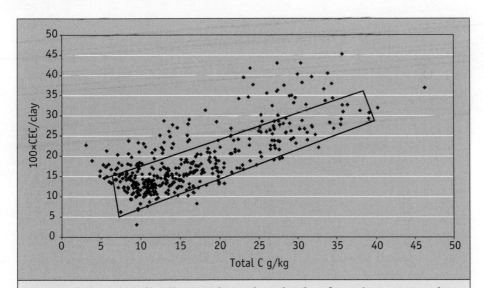

Figure 2. Scatter plot for all CEC/clay and total C data from the Amazonia long-established land use systems. The delimited zone contains the set from flat sites; sloping sites scatter throughout.

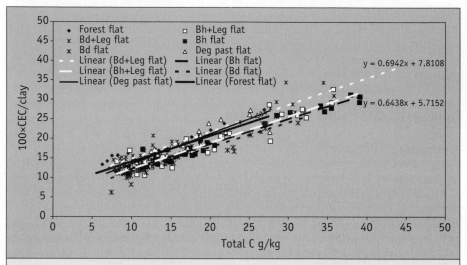

Figure 3. Scatter plot for CEC/clay and C data from Amazonia long-established systems on flat sites.

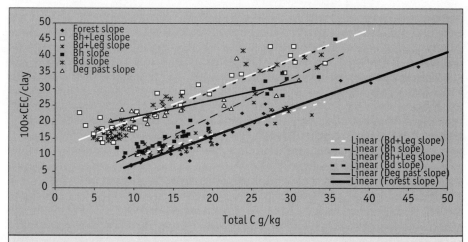

Figure 4. Scatter plot for CEC/clay and C data from Amazonia long-established systems on sloping sites.

Table 1. Regressions between 100×CEC/clay and total C for different long-established land use systems in Amazonia. Outliers removed. 100 × CEC/clay = a × C + b

Position	Land use	a	b	r²
Flat	Forest	0.66	7.44	0.84
	B. humidicola	0.64	5.71	0.95
	B. decumbens	0.64	5.09	0.88
	B. humidicola + legume	0.67	5.40	0.89
	B. decumbens + legume	0.69	7.81	0.89
	Degraded pasture	0.77	5.80	0.88
Sloping	Forest	0.85	-1.35	0.92
	B. humidicola	1.15	-1.28	0.94
	B. decumbens	0.94	10.59	0.90
	B. humidicola + legume	0.91	11.65	0.88
	B. decumbens + legume	0.69	2.64	0.85
	Degraded pasture	0.51	16.44	0.55

Table 1 shows that, within the margin of error:
1. in flat positions, the OM under all land uses has a similar CEC per gram (parameter a);
2. in flat positions, the cut off on the Y axis (parameter b) is only marginally different for all land uses, indicating similar soils and a homogeneous data set;
3. in sloping positions the CEC of the OM under some land uses is significantly higher, indicating a different OM quality;
4. in sloping positions, there is a considerable difference in the cut off on the Y axis, indicating significant differences in soils that may impede comparison between land uses (a negative cut off may indicate that the clay fraction has hardly any, which may occur in iron-rich oxisols;
5. in sloping positions, the correlation for degraded pasture is rather low, which may explain the different slope and cut off. Erosion, resulting in the exposure of subsoil layers, may be the cause of variation.

The cut-off on the Y axis of the regression lines in Figure 3 indicate that the soils of the long-established land use systems in sloping positions belong to two groups. The forest soils, those of 'B. decumbens+legume' and those of 'B. humidicola' have a cut-off close to zero, indicating oxisols virtually without CEC in the clay fraction. The other group, containing '*B. humidicola+legume*', '*B. decumbens*' and '*Degraded pasture*' have a cut-off of 10 or larger and thus have clays with considerably higher CEC. These soils probably belong to ultisols. Indeed, the extrapolation studies of Van Engelen (Chapter 10, this volume) indicate two major soil groups in this area.

CEC data of the newly established land use trials (Figure 5) indicate that the soils of the sloping and flat topographies are fairly homogeneous. As with the soils of the long-established land use systems, there is a clear difference between the two, while also the OM matter in the two sets appears to have different properties (different slope of the regression line).

BD data also indicate considerable difference between the flat and sloping areas. Figure 6 gives BD data of the new land use trials in Amazonia. The values have been plotted in the middle of each sampled layer, but the values for the sloping sites have been slightly displaced to improve clarity. The flat topography soils have lower mean BD than the sloping topography. Moreover, the range in BD of the sloping topography is higher. As the BDs in the top soils of sloping sites reach values as high as 1.4 and 1.6 kg/dm³, this suggests that some of the soils are

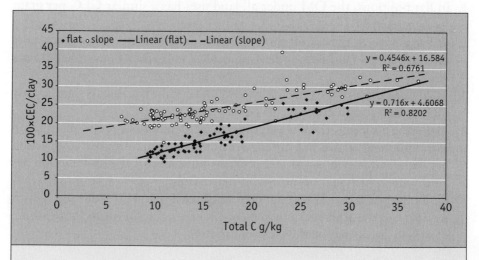

Figure 5. Scatter plot for CEC/clay and C data from Amazonia short-term experiments on flat and sloping sites.

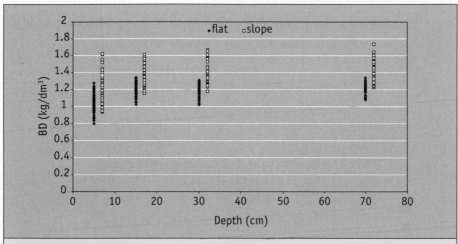

Figure 6. Bulk density data of soils on flat and sloping sites; new land use experiments, Amazonia.

eroded. This may be an additional factor of variation for the CEC data, because clay mineralogy, and thus its CEC, may change with depth.

Andean Hillsides

The Andean Hillside soils show considerably more variation than those from Amazonia. Both the first and the second sampling show a large scattering of data and, except for the forest plots, rather low correlations between 100×CEC/clay and total C (Table 2). The results for Dagua and El Dovio are presented in Figures 7 and 8. In these figures, only correlations with $r^2 > 0.5$ are shown.

For Dagua, significant correlations between 100×CEC/clay and C-total were found for the forest plots, the degraded pasture, and the degraded soil. The two samplings of Forest plot 2 are significantly different (which may be a systematic laboratory error), and both are significantly different from Forest plot 1. Considering that the vertical scale of the plot is 6 times that of the Amazonia plots, there is considerable variation in soil properties, especially in the CEC of the clay fraction. This variation may be due to stratification of the volcanic parent material, which affects both the clay mineral assemblage and the OM decomposition (buried soils). The OM in the two forest plots appears to have a different quality (strongly different slope of the correlation). Most treatments appear to have similar soils to Forest plot 2.

Table 2. Regressions between 100×CEC/clay and total C for long-established land use systems of the Andean hill sides. Outliers removed.
$100 \times CEC/clay = a \times C + b$.

Location	Land use	a	b	r^2
Dagua	Forest 1	1.67	48.60	0.86
	Forest 2-1st sampling	0.41	11.88	0.56
	Forest 2- 2d sampling	0.58	29.21	0.47
	Degraded pasture 1st sampling	1.17	6.97	0.89
	Degraded pasture 2d sampling	1.02	7.65	0.88
	Improved pasture 1st sampling	Data removed, see later		
	Improved pasture 2d sampling	0.53	30.40	0.32
	Forage bank 1st sampling	-0.15	89.43	0.01
	Forage bank 2d sampling	-0.57	103.6	0.05
	Degraded soil	1.12	10.46	0.62
El Dovio	Forest 1st, sampling	2.44	26.51	0.74
	Forest 2d sampling	2.65	21.14	0.73
	Degraded pasture 1st sampling	0.88	67.95	0.74
	Degraded pasture 2d sampling	1.36	65.58	0.73
	Improved pasture 1st sampling	0.98	45.22	0.33
	Improved pasture 2d sampling	0.88	56.40	0.31
	Forage bank 1st sampling	-0.76	106.0	0.10
	Forage bank 2d sampling	-1.31	111.0	0.39

In El Dovio, the two samplings of the forest plot gave similar results (Figure 8), but the scatter within the data set indicates considerable soil variation. The CEC of the parent material (Y-axis cut-off) is higher for El Dovio than for Dagua.

In both El Dovio and Dagua, the mixed forage bank plots show the largest scatter with the worst correlations. In Dagua, they are more related to the Forest plot 1 than to any others. In El Dovio, the values are scattered throughout. In the mixed forage banks alone, there is no positive relation between OM contents and CEC, which might indicate severe degradation of the soil and loss of active humus fractions.

The newly established experiments in Dagua and El Dovio show a similar picture (Figure 9). The Dagua site appears adequately homogeneous, but the data for

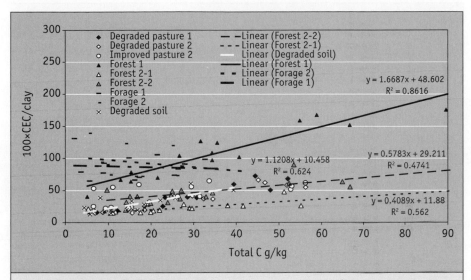

Figure 7. Relations between 100×CEC/clay and total C for long-established land use systems of Dagua.

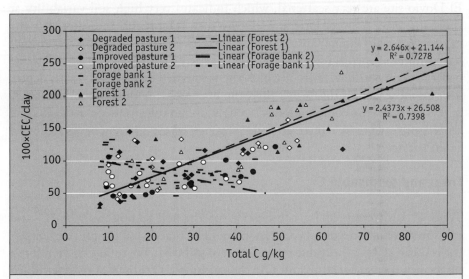

Figure 8. Relations between 100×CEC/clay and Total C for long-established land use systems of El Dovio.

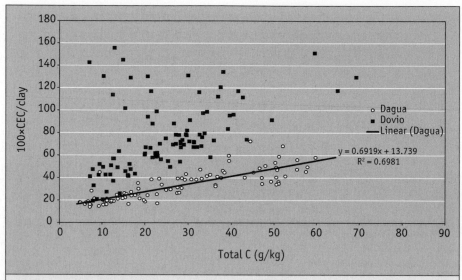

Figure 9. Relations between 100×CEC/clay and Total C for short-term land use experiments in Dagua and El Dovio.

El Dovio site are very scattered, which indicates very high soil variability at short distance. A closer look at El Dovio data (Figure 10) indicates that the experiments with a monoculture of *Brachiaria*, *Brachiaria+A.*, and *Regeneration* have similar soils with high correlations between CEC/clay and total C, while other experiments have soils of which the properties strongly vary with depth. This implies that only the former three can be compared. The extreme variability at this site is due to a combination of strongly layered volcanic deposits (including buried soils) and erosion on steep slopes, through which different layers are found at the surface.

Total and oxidisable C

Total C is determined by combustion of the soil sample at 900 °C and measurement of the evolved CO_2. In soil samples that do not contain carbonates the resulting CO_2 is fully due to oxidation of soil OM. The total C is therefore a direct measurement of all organic C in the soil sample. Oxidisable C is determined by wet oxidation of a soil sample and back-titration of the remaining oxidant. The result is an amount of oxidant used, which is converted to organic C using an assumed recovery factor of 76% (natural variation 60-86%; SSAC, 1996). If organic C is converted to soil OM, a mean value of 58% C in soil OM is used.

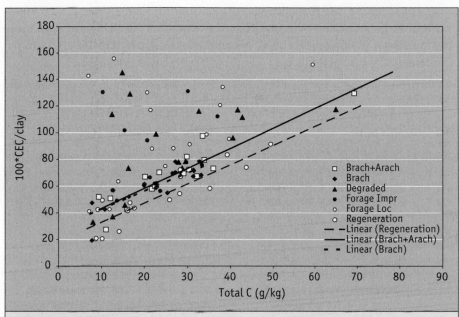

Figure 10. Relations between 100×CEC/clay and Total C for newly established land use experiments in El Dovio.

Also this factor is variable. The recovery factor of the wet chemical procedure depends, among others, on the kind of OM. Charcoal is inefficiently oxidised by wet chemical methods, and also clay-OM bonds may offer a certain protection to wet oxidation, while some cation associations may have a similar effect.

In most soils, a linear regression is expected between both C determinations. Deviations from a 1:1 ratio may be due to a combination of the assumed efficiency factor and the assumed C content of oxidisable matter. Therefore, the correlations have not been forced through the origin (0,0). The cut-off on the Y axis is an indication of the adequacy of these conversion factors. Fully adequate conversion factors for wet oxidation would result in zero cut-off. The cut-off is therefore not considered, and the coefficient a (slope) of the regressions in Tables 3 and 4 therefore indicates the fraction of oxidisable C with respect to total C.

In addition to the effect of conversion factors, there may be real differences between total and oxidisable C, due to the presence of clay organic complexes and charcoal. For this project, it was assumed that the difference between total C and oxidisable C might give some indication of the less-oxidisable humus

Table 3. Regressions between total C and oxidisable C for different land uses and positions in long-established systems of Amazonia. Outliers removed. Oxidisable C = a × total C + b

Position	Land use	a	B	r^2
Flat	Forest	0.93	-4.39	0.97
	B. humidicola	0.86	-4.26	0.93
	B. decumbens	0.86	-4.26	0.93
	B. humidicola + legume	0.94	-4.91	0.98
	B. decumbens + legume	0.86	-4.26	0.93
	Degraded pasture	0.86	-4.26	0.93
Sloping	Forest	0.81	-2.57	0.97
	B. humidicola	0.97	-6.36	0.97
	B. decumbens	0.97	-6.36	0.97
	B. humidicola + legume	0.91	-2.67	0.98
	B. decumbens + legume	0.97	-6.36	0.97
	Degraded pasture	0.97	-6.36	0.97

fraction. Erratic behavior of the relation between the two may indicate either analytical errors or a significant admixture of non-oxidisable C. In both cases, it is better to eliminate such samples before statistical treatment of the results. Native forest plots do not usually have charcoal, and therefore these can be used as a reference.

Amazonia

An analysis of the forest plots (Figure 11) shows that both for flat and sloping sites, there is a high correlation between oxidisable C and total C. Two data points of the slope positions deviate considerably from the general trend, which suggests that either an analytical error was made or the sites are not representative for the whole. If such points are eliminated, data sets of all treatments on flat and sloping sites are very close together and statistically similar (Table 3). This is an ideal situation for comparison of land use effects.

Soils under B. decumbens show a high correlation in flat positions and considerably more scatter in sloping positions (Figure 12). In the previous section it was shown that the soils of this treatment differed considerably between flat and sloping

Table 4. Regressions for oxidisable C and total C for different land use systems and samplings in long-established Andean Hillside plots. Outliers removed. Oxidisable C = a × total C + b

Position	Land use	a	b	r^2
Dagua	Forest 1	0.81	+0.01	0.96
	Forest 2 – 1st sampling	0.97	-3.21	0.99
	Forest 2 – 2nd sampling	0.80	-1.67	0.99
	Improved pasture, 1st sampling	0.69	2.99	0.61
	Improved pasture, 2nd sampling	0.87	-1.90	0.98
	Degraded pasture, 1st sampling	0.92	-3.92	0.96
	Degraded pasture, 2nd sampling	0.87	-4.90	0.98
	Mixed forage bank, 1st sampling	0.88	-1.67	0.99
	Mixed forage bank, 2nd sampling	0.97	-4.39	0.98
	Degraded soil	0.98	-3.84	0.97
El Dovio	Forest, 1st sampling	0.85	-2.12	0.97
	Forest, 2nd sampling	0.81	-0.91	0.96
	Improved pasture, 1st sampling	0.88	-2.97	0.98
	Improved pasture, 2nd sampling	0.87	-5.30	0.97
	Degraded pasture, 1st sampling	0.84	-0.96	0.95
	Degraded pasture, 2nd sampling	0.84	-3.07	0.99
	Mixed forage bank, 1st sampling	0.78	-0.45	0.96
	Mixed forage bank, 2nd sampling	0.85	-4.80	0.98

sites, which may be the cause of the scatter. Because of the large scatter, there are no obvious error points, but the whole data set of the sloping sites was excluded from further evaluations. *B. decumbens* + legume gives very similar relations to those of *B. decumbens* alone (Figure 13), but without differences between flat and sloping sites.

Under both *B. humidicola* and *B. humidicola* + legume, the regressions for flat and sloping sites have a similar slope, but a significantly different cut off (shown in Figure 14 for *B. humidicola*). This suggests that under *B. humidicola*, sloping sites have relatively more oxidisable C than flat sites. This remains unexplained so far, but it is possible that microbial organic matter, which is part of the oxidisable C, constitutes a larger fraction in these well-drained soils.

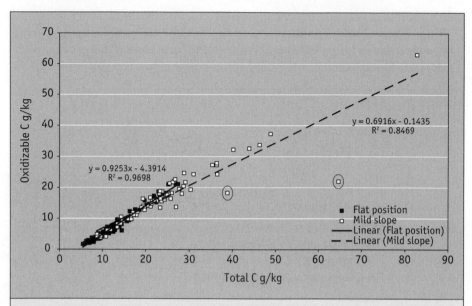

Figure 11. Oxidisable C and total C in Amazonia forest plots. Circles indicate outliers.

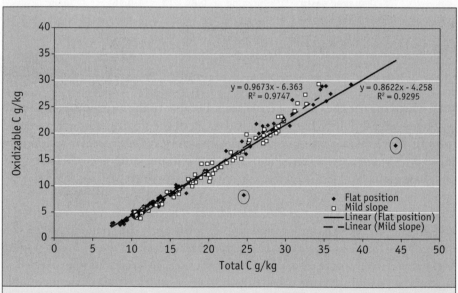

Figure 12. Oxidisable C and total C in long-established land use systems in Amazonia under *B. decumbens*.

Carbon sequestration in tropical grassland ecosytems

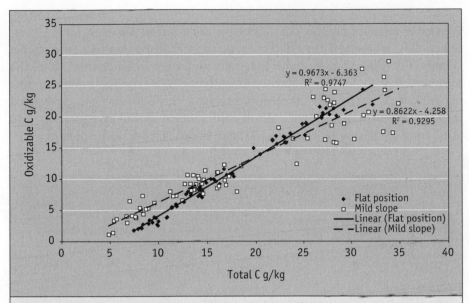

Figure 13. Oxidisable C and total C in long-established systems in Amazonia under *B. decumbens* + legume. Circles indicate outliers.

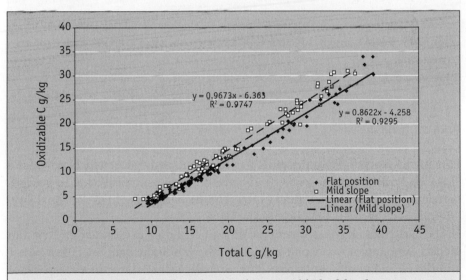

Figure 14. Total C and oxidisable C in long-established land use systems in Amazonia under *B. humidicola*.

A similar vegetation-dependent effect on the relation between total C and oxidisable C is observed in the data of the newly estabalished land use experiments (Figure 15). The graph indicates that soils under the 'regeneration' plots, the 'degraded pasture' and the 'improved forage bank' have relatively lower amounts of oxidisable C than under the other vegetations. This difference was established during the 3.3 years of the trial. Similar differences induced by vegetation were also found in the new plots on sloping land, but the trends were different.

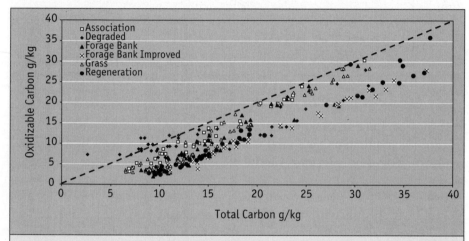

Figure 15. Total C and oxidisable C in newly established land use experiments, Amazonia.

Andean Hillsides

Part of the Andean Hillsides plots has been sampled twice, which provides a check on both plot homogeneity and consistency of analytical data. In general, correlations are very high indeed (Table 4), which indicates high data quality.

In the data from Dagua, a systematic difference shows up between the first and second sampling of forest plot 2 (Figure 16). While the data for Forest plot 1 and the second sampling for plot 2 are virtually identical, the first sampling of plot 2 is different. This is probably due to a systematic difference (batch effect) in laboratory analyses.

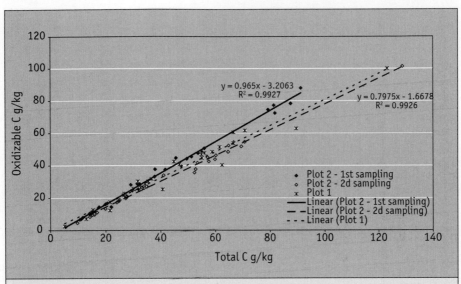

Figure 16. Total C and oxidisable C in forest plots of the long-established land use system in Dagua.

The first data set of the improved pasture in the long-established land use system in Dagua, shows considerable scatter while the second one offers a high correlation between oxidisable C and total C (Figure 17). Because the second data set is a re-sampling of the same plots, it is likely that the scatter in the first set of laboratory analyses is due to errors, and the set was therefore eliminated.

In the degraded pasture in the long-established land use system in Dagua, data from the first and second sampling are virtually similar and only one outlier was spotted. The 'mixed forage bank' plots and the degraded soil plots from Dagua have well correlated and consistent data without outliers (Table 4).

The first and second sampling of the El Dovio forest plots show a homogeneous data set with only two obvious outliers (Figure 18, outliers removed). The data from long term improved pasture, degraded pasture, and mixed forage bank plots suggest a slight systematic difference in laboratory results (Table 4). Only the degraded pasture data showed three outliers.

The C data of the newly established land use experiments in Dagua probably contain a systematic laboratory error (Figure 19). Above a total C content of about 40 g/kg, the whole data set shifts to the right but the slope of the

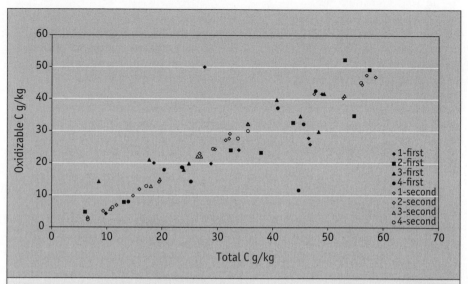

Figure 17. Total and oxidisable C in improved pasture plots of the long-established land use systems in Dagua. Points 1-4: sloping positions; First/second: sampling stage. Black symbol: unreliable data.

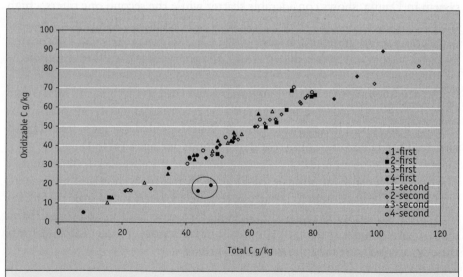

Figure 18. Total and oxidisable C in El Dovio long-established forest plots. Points 1-4: slope positions; First/second: sampling phase. Outliers encircled.

correlation was maintained. At lower total C contents, a few samples appear to follow this trend. This is probably the effect of a batch of samples that was analysed with differently made chemical solutions. In this case it is prudent not to use the oxidisable data.

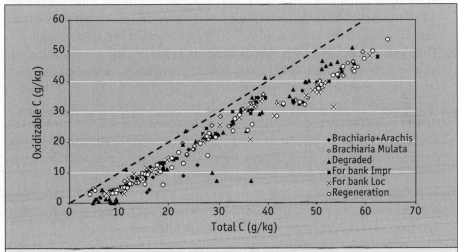

Figure 19. Total and oxidisable C in the newly established land use experiments at Dagua

Conclusions

The methods of data evaluation used here allow the following conclusions concerning soil homogeneity and quality of analytical data:

1. The long-established flat topography plots in Amazonia are homogeneous with respect to soils and the OM quality is similar.
2. The differences in CEC$_{clay}$ (Y axis cutoff) of the slope soils from Amazonia suggest the presence of two soil groups. OM quality is not widely different between the two groups and it is unlikely that the differences will have caused essential differences in C stocks.
3. Only under the long-established *B. humidicola*, in Amazonia, slope appears to increase the amount of oxidisable C. This may be due to increased production of microbial biomass.

4. The newly established land use experiments in Amazonia were established on homogeneous soils, both on flat and sloping sites.
5. The soils of the Andean Hillsides, both in Dagua and El Dovio, show considerable differences (as would be expected in high altitude volcanic soils). The apparent stratification of the soils induces differences in characteristics of the total C fraction, but it is not clear whether this might affect the budget calculations. In both Dagua and El Dovio, the soils under 'mixed forage banks' show aberrant behavior of total C, which may indicate soil degradation.
6. The fields of the newly established land use trials in Dagua are sufficiently homogeneous. In those of El Dovio, however, variability is so high that only selected treatments can be compared.
7. In most soils, there is a high correlation between oxidisable C and total C. If, within one kind of land use, the values of both parameters show considerable scatter, this suggests laboratory errors. Such data sets, together with outliers in otherwise well-correlated data, should be removed from the data set to improve the reliability of C stock comparisons.
8. Repetition of sampling on the same plots allows the detection of systematic differences in laboratory data. Small systematic differences were found in various cases, and only one set of data was invalidated.
9. The removal of erroneous data as suggested by the present analysis considerably improves the data set, which is essential for C stock calculations and comparisons.

From this we may draw a number of more general conclusions:
10. Despite the expected natural soil variability in long-established pasture and silvopastoral systems in commercial farms, the project selected sufficiently homogeneous plots for both the flat and sloping positions in Amazonia, and as-homogeneous-as-natural conditions-permit, in the eroded Andean Hillsides of Colombia.
11. It is virtually impossible to find homogeneous soils in strongly sloping areas, especially on very small farms of multi-crop subsistence agriculture, as those on eroded Andean hillsides.
12. Soil variability is greater in sloping areas than in flat topographies. It cannot be avoided, and the challenge for any type of field based research is to understand it, quantify it and use it properly to reach valid comparisons and conclusions.
13. According to the chosen experimental design only El Dovio experimental data need to be analysed by eliminating field or treatments with deviating soils. The information thus obtained can be extrapolated to areas with similar slope, soil, and agroclimatic conditions.

Chapter 5. Factors affecting soil C stocks: a multivariate analysis approach

P. Buurman, M.C. Amézquita and H. F. Ramírez

Introduction

From the analysis presented it is clear that 'land use system' is the most important factor determining soil C stocks. In the Andean Hillsides ecosystem 'land use system' and 'slope gradient', accounted for 60% and 11% of the variability in total C stocks, respectively, leaving a non-explained variability, excluding factor interactions, of 24%. In the tropical forest ecosystem, Amazonia, flat topography, Colombia 'land use system' and 'spatial variability within the same land use system' accounted for 43% and 10% of the variability in total C stocks, respectively, leaving a non explained variation of 47%. In the sloping topography, Amazonia, the same two factors, together with 'slope gradient', accounted for 74%, 7%, and 7% of the total variation, respectively, with a remaining non explained variation of 12%. These findings indicate that there are other unknown factors affecting soil C stocks that should be studied to have a more complete explanation of soil C behaviour.

From the data presented in Chapter 4 on soil variabilty and data consistrency and from theoretical considerations it is clear that the variables total C, oxidisable C and CEC are strongly correlated. From these data, however, it is not possible to draw conclusions as to possible relations between C contents and other soil variables. Similarly, the effect of land use system on these relations combined remained unexplored.

The aim of this Chapter is to identify other factors, apart from land use, that affect soil C stocks. The aim is to come to an understanding of the relationships between soil properties and C and N contents at a given site and how these relationships are affected by the land use system.

Principal Component Analysis (PCA) is applied to find relations between the apparently independent soil variables and their dependence on land use system. PCA is a multivariate statistical method used for reduction of dimensionality of a problem. PCA starts by calculating the correlation matrix between the original set of variables and based on the correlation structure it estimates a new set of independent factors, called *Principal Components* or *Factors*, which are linear combinations of the original variables. The maximum number of Factors used in PCA equals the number of original variables. In that case, all variation in the individual variables is explained by the total of factors. In practice, however, a far smaller number of (uncorrelated) factors is used to explain a high proportion of the variability present in the original variables. Even in our data sets, where the number of variables is only seven, a reduction to two or three explaining factors helps to analyse relations between variables and to find general shifts in such relations due to land use.

Four sites were analysed: Dagua and El Dovio in the Andean Hillsides, Colombia, and flat and sloping topography sites in the humid tropical forest ecosystem of Amazonia, Colombia. For each site a separate analysis of the 0-10 cm and 20-40 cm layers was made. (The analysis for Costa Rica is reported in Chapter 3).

The original variables used for PCA were: total C, oxidisable C, total N, cation exchange capacity (CEC), BD, clay content and sand content. Table 1 lists the maximum and minimum values of the seven variables for the different sites and soil depths.

The variables sand and clay content are largely independent of land use system, although some land uses may accelerate removal of fine material from the topsoil. The other variables may change with land use. Large ranges of sand and clay contents within one data set indicate different parent materials. In Amazonia

Table 1. Minimum and maximum values of the original variables.

Site	Depth	BD	C tot	C oxid	N tot	CEC	Clay	Sand
Amazonia flat	0-10	0.5-1.4	1.3-4.8	1.0-3.4	0.1-0.4	6-19	27-49	24-55
	20-40	0.9-2.1	1.0-2.6	0.3-1.2	0.1-0.2	5-17	31-68	16-41
Amazonia sloping	0-10	0.6-1.5	1.8-4.9	1.2-3.4	0.1-0.4	6-18	27-48	2-66
	20-40	0.8-1.7	0.6-2.5	0.3-1.4	0.0-0.2	5-14	35-67	0-49
Dagua	0-10	0.4-1.2	1.2-12.9	0.6-6.3	0.1-1.3	10-36	19-72	7-67
	20-40	0.6-1.4	0.6-5.3	0.2-3.0	0.0-0.5	8-32	20-72	5-62
El Dovio	0-10	0.4-1.2	2.2-11.3	2.3-6.9	0.2-1.0	14-42	17-33	33-64
	20-40	0.6-1.5	0.6-8.1	0.8-4.2	0.0-0.5	11-32	17-52	22-57

sloping topography, for example, the low sand contents are restricted to the degraded pasture. In Dagua, low sand contents are found in the degraded soil, while low clay contents are found in one of the forest systems (data not shown).

Results and discussion

In 7 out of the 8 cases analysed (4 sites x 2 soil depths) only two significant Factors (with Eigen values >1) accounted for 62 to 80% of the total variation in the original variables. Only for the subsoil of El Dovio, three significant Factors (explained variation = 88%) could be extracted (Table 2). The remaining variation of 12-38% remains unexplained and should be considered as random variation due to field variability and analytical errors. The amount of unexplained variation is not related to the depth of the layer. It is striking that the lowest explained variation is encountered in the Amazonia data sets, which are most homogeneous with respect to soils (see Chapter 2.3). Because in 7 out of the 8 cases studied only two factors have Eigen values that are significant, we will describe all cases by these two factors alone.

Table 2 also lists the loadings of the variables on each Factor (or Principal Component) scores. These loadings which are depicted for El Dovio (0-10 cm) in Figure 1 and indicated by relative positions in Figures 3-6, are essential for understanding differences between sample sets in the Factor scores plots. This will be explained below.

The loadings of variables on a Factor may be positive in one data set and negative in the next one. This is not essentially different; a data matrix in principal component analysis may 'flip' without significant change in content. Therefore, Factor 1 may show either highly positive *or* highly negative loadings for total C, oxidisable C and total N.

In all cases, the variables total C, oxidisable C and total N cluster together in the space defined by Factors 1 and 2, indicating their mutual dependence. The example of Figure 1 is from Dagua (0-10 cm depth).

In Figure 1 the two variables that are in the part of factor space opposite to C and N contents, are BD and clay content. These are negatively correlated with the variables on the right. Such a relation can be explained as follows: higher clay contents tend to cause a larger BD, while high contents of organic matter (total C) decrease BD through two effects: (1) organic matter has a lower particle

Table 2. Factor loadings on- and explained variance by- the first two Principal Components (Factor 1 and Factor 2) for each site and soil depth.

Site and soil depth	Amazonia flat				Amazonia sloping				Dagua				El Dovio				
	0-10 cm		20-40 cm		0-10 cm		20-40 cm		0-10 cm		20-40 cm		0-10 cm		20-40 cm		
Factors and their explained variation (%)	F1	F2	F1	F2	F1	F2	F1	F2	F1	F2	F1	F2	F1	F2	F1	F2	F3
	45.3	26.6	47.3	14.3	37.0	26.2	48.2	25.1	55.9	23.7	47.5	26.7	60.7	15.3	51.2	20.9	15.6
Factor loadings on original variables																	
Total C	-0.86	-0.21	-0.80	-0.29	-0.70	0.18	-0.75	0.54	0.80	0.47	0.66	0.60	0.86	0.01	0.81	-0.47	-0.26
Oxid C	-0.87	-0.06	-0.70	-0.33	-0.73	-0.28	-0.77	0.40	0.83	0.21	0.74	0.23	0.92	0.04	0.84	-0.30	0.01
Total N	-0.81	-0.27	-0.73	-0.31	-0.84	-0.01	-0.53	-0.14	0.76	0.44	0.79	0.42	0.79	-0.06	0.80	-0.45	-0.28
CEC	-0.63	0.54	-0.75	0.13	-0.66	-0.60	-0.55	-0.76	0.38	-0.78	0.30	-0.74	0.82	0.31	0.70	0.26	0.53
Clay	-0.21	0.86	-0.71	0.13	0.46	-0.70	-0.85	-0.08	-0.69	0.60	-0.75	0.61	-0.24	0.93	-0.59	-0.72	0.27
Sand	0.05	-0.83	0.46	-0.79	-0.15	0.93	0.66	0.71	0.83	-0.43	0.78	-0.53	0.78	-0.25	0.68	0.60	-0.27
BD	0.76	0.14	0.60	-0.23	0.43	-0.05	0.69	-0.44	-0.83	-0.21	-0.68	-0.30	-0.84	-0.20	-0.54	-0.03	-0.72

Carbon sequestration in tropical grassland ecosytems

density than mineral matter, and (2) the presence of organic matter stabilises porous structures, which have a lower BD. Further, clay and sand contents are found in opposing locations, suggesting that the complementary silt fraction (clay+sand+silt = 100%) does not show much variation. Although a positive relation between clay and organic matter contents is sometimes found, especially in soils of temperate areas, this is not the case here. The influence of clay on BD appears to be more important than its effect on organic matter content.

In the data sets of Amazonia flat topography (both depths) and Amazonia sloping topography (20-40 cm), however, the sand fraction instead of the clay fraction plots at the same side as BD. This may be due to very low sand contents in part of the data set, but a definite explanation is not available. Clay and sand still plot at opposite sites of the diagram.

CEC_{soil} depends on both clay content and organic matter. In the factor space of Figure 1 it does not cluster with any of these variables, nor between them, suggesting that other factors than the bare contents of clay and C, such as

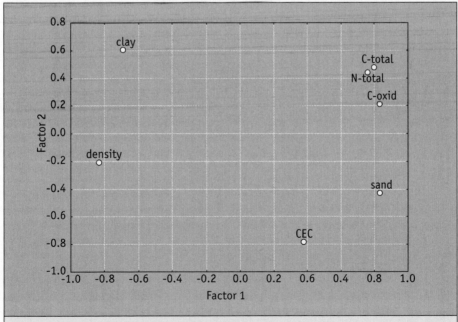

Figure 1. Loadings of seven variables in the space defined by Factor 1 and 2. Data set Dagua (0-10 cm depth).

different CEC of the clay fraction and different humification degree of organic matter, may influence CEC_{soil}.

Figure 2 shows the plot of all sampling points for Dagua (0-10 cm depth) in F1F2 space. Points belonging to the same land use system are grouped. In Figure 2, the various land use systems occur in different areas of the factor space, which means that both the level of variables and the relations between them vary with land use. The two forest plots do not overlap, but occur both towards the right. This implies that they have higher C, N and sand contents, and CEC than the other land uses. Both the improved and the degraded pasture plot in the center, indicating that the improvement of pasture did not affect the relation between the variables and that the soils are similar. The degraded soil occurs rather to the left, indicating both higher BD and lower C contents. When the forage bank is compared with both pastures, it appears that the shift is on Factor 2 and hardly

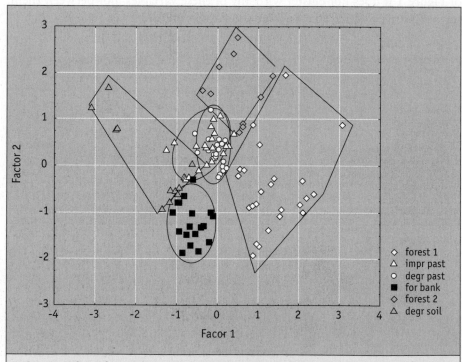

Figure 2. Plot of samples in F1F2 space, grouped by land use system. Dagua, 0-10 cm depth.

Carbon sequestration in tropical grassland ecosytems

on Factor 1. This implies a loss of both clay and organic C (both at the top of the diagram), but no change in BD.

For the subsoil of the Dagua systems (not shown), forest plot 1 remains separated and the forage bank still occur at the bottom, but the pastures, the degraded soil and the second forest plot overlap, suggesting that the land use effect on the variables – at this time scale – does not reach depths of 20-40 cm.

For El Dovio area, part of the land use systems were not included in the analysis because the soil variation appeared too large (see above). The scores plot of El Dovio (0-10 cm) samples in F1F2 space is given in Figure 3. Instead of presenting a separate plot of Factor loadings and Factor scores, as in Figures 1 and 2, the relative position of the variables in factor space have been indicated in this scores plot.

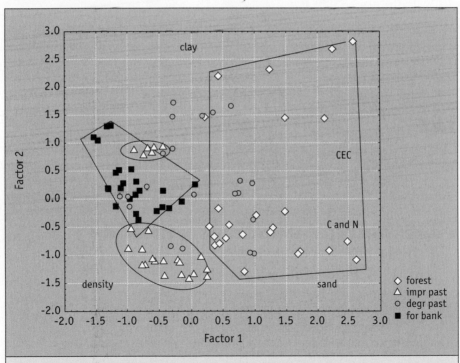

Figure 3. Plot of samples in F1F2 space, grouped by land use system, El Dovio, 0-10 cm depth.

Factor 2 of Figure 3 illustrates some of the remaining heterogeneity of El Dovio land use systems. The degraded pasture overlaps with all other systems. The improved pasture appears to have a somewhat higher BD (bd at lower left) but falls apart into two clusters. The forage bank is the most homogeneous unit. At a depth of 20-40 cm, El Dovio samples – apart from the forest – similarly show a large overlap between systems. The third Factor, which explains about 16% of the variation in El Dovio (20-40 cm) samples, is mainly related to clay content and sets apart some samples of the degraded pasture (not shown).

Because of their very similar soils, the plots of the topsoil (0-10 cm) samples of land use systems in Amazonia flat topography largely overlap (Figure 4). Only the groups 'forest 1', '*B. humidicola monoculture*' and '*B. humidicola + legumes*' have been delineated. The plot of the samples in F1F2 space is largely determined by the four extremes: C and N, BD, clay, sand. In this system, where soils and humification are homogeneous, CEC occurs between its contributors total C

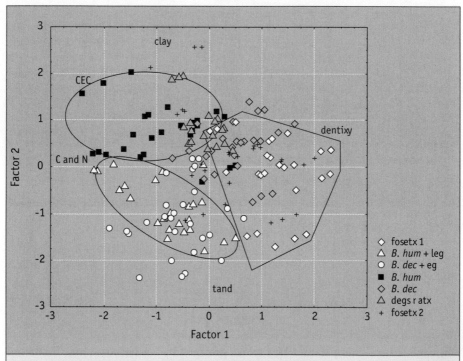

Figure 4. Plot of samples in F1F2 space, grouped by land use system. Amazoniam flat topography, 0-10 cm depth.

and clay. Within this rather homogeneous data set, differences nevertheless show up. Shifts on the Y axis should be largely due to original variation in soil properties (sand versus clay contents), while shifts on the X axis are related to changes caused by the land use system. Figure 4, therefore, suggests that the difference between, e.g. *B. humidicola* monoculture (closed squares) and *B. humidicola + legumes* (open triangles) is largely due to original differences in soil, while the difference between *B. humidicola* monoculture and *forest* (crosses and open diamonds) should be largely due to the land use itself.

The plot of the subsoils (20-40 cm) of Amazonia flat topography(not shown) shows largely the same picture.

Although the soils of the sloping areas in Amazonia were less homogeneous, the plot of samples in F1F2 space, for the top layer (Figure 5) shows largely overlapping populations. Considering the underlying factor loadings, the vertical

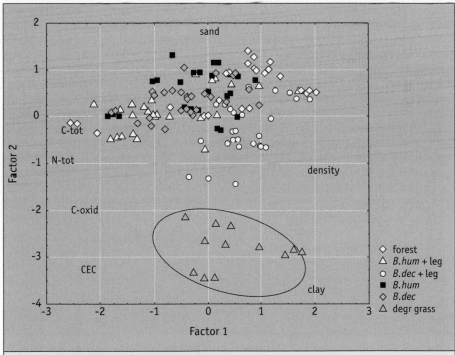

Figure 5. Plot of samples in F1F2 space, grouped by land use system. Amazonia sloping topography, 0-10 cm depth.

differentiation between the degraded grassland and the other land uses is mainly a matter of texture.

The diagram of the subsoils (20-40 cm) is essentially different (Figure 6). The underlying factor loadings indicate that shifts along the F1 axis are caused by land use (density-C shifts) and texture (clay-sand), while those on the F2 axis depend on texture and CEC. Subsoils fall apart into two groups, with different CEC of the clay fraction, which is reflected here. The degraded pasture, which is most strongly eroded, shows the higher CECs of the surfacing subsoil. Both *B. decumbens* in monoculture and *B. humidicola* with legumes appear to have higher BD and lower C and N contents than the other land use systems. The forest, with highest C contents and lowest CEC probably represents the least eroded soils.

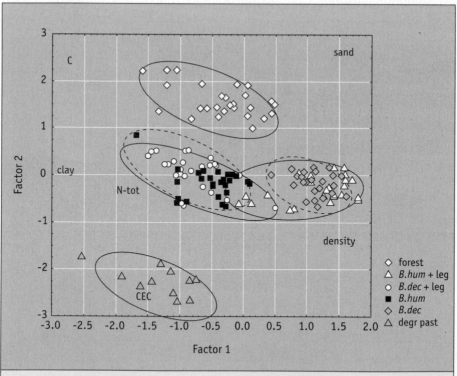

Figure 6. Plot of samples in F1F2 space, grouped by land use system. Amazonia sloping topography, 20-40 cm depth.

Conclusions

1. As expected, the variables total C, oxidisable C and total N are strongly correlated in all samples studied. Increasing C contents cause a decrease in BD, so that this variable is found at the opposite side of factor space. Although CEC is built up of contributions of organic matter (total C) and clay, this relation is not always apparent. Causes for deviations are heterogeneous parent materials (different CEC of the clay fraction) and different humification degree of the organic fraction. Because silt fractions are rather constant, clay and sand contents are found at opposite sides of factor space.

2. Explained variation of the seven variables by two factors ranges between 62 and 80%. Factor 1, which explains 37-61% of the variation, is mainly the effect of land use on C stocks and BD, while Factor 2 largely explains variability caused by variations in parent material. If land uses show a separation on Factor 2, this implies that the plots were not completely similar. This is most clearly seen in the subsoils of the Amazonia sloping topography data set. Because subsoil C stocks in both soils were largely similar, this effect does not interfere with the conclusions regarding effect of land use on C stocks.

3. From the present analysis it can be concluded that, although the land use system is the most important factor determining soil C stocks, some relations between soil variables are affected by land use, and changes in such relations also influence the C accumulation capacity of soils.

Chapter 6. Methodology of socio-economic research

J. Gobbi, B.L. Ramírez, J. Muñoz and P. Cuellar

Introduction

Given the growing concerns with global warming and the social and economic importance of livestock in Latin America, this project aims to identify livestock production systems that are able to capture C and are at the same time financially attractive to farmers.

Questions addressed by the socio-economic research are:
1. How much C can different livestock systems in the ecosystems covered by this project store?
2. Are those land use systems financially viable *vis-à-vis* current land use practices?
3. What is the potential financial effect of a payment for C sequestration on farm income and on the financial feasibility of land use systems that sequester C?

The objectives of the socio-economic research were to:
- characterise the socio-economic conditions of farms participating in the project;
- describe land use on farms included in the project;
- define establishment costs and operating expenditures of different C sequestering livestock systems, as well as their production and revenue levels;
- explore the financial feasibility of investing in different C sequestering livestock systems;
- develop models on the financial effect that a potential payment for C storage may have as an incentive to incorporate C sequestering livestock systems on farms;
- provide policy guidelines to promote the implementation of C sequestering livestock systems by farms in Tropical America.

J. Gobbi, B.L. Ramírez, J. Muñoz and P. Cuellar

Conceptual framework of the socio-economic research

The socio-economic questions posed by the project require a conceptual framework based on a systems approach. The conceptual framework is developed to establish the relations between the farms in their external context in order to help identify relevant characteristics of the system under study that may influence the implementation of C sequestering livestock systems. Accordingly, this framework provides conceptual unity to the study and allows to link socio-economic results obtained at different levels. Therefore, the farm, which is a mosaic of land uses with different capacities to sequester C, is part of a framework that encompasses local, national and international scales, that exert external influences on the farm.

From this framework, it is understood that processes and variables operating at a higher hierarchical scale have a direct effect on the processes and variables operating at lower scales, but not *vice versa*. For example, the structure of incentives, both positive and negative, directed to the livestock sector implemented by national governments favours the adoption of certain livestock production systems above others. In this sense, decisions taken by farmers are constrained not only by their availability of production factors (i.e. capital, land and labour) but also by a variable (incentive structure) that operates at a scale higher than the farm and imposes a direct influence on farmer's ability to plan in the long term.

Throughout the application of the conceptual framework, socio-economic research studies the financial feasibility of investing in C sequestrating livestock systems as well as identifies policy elements and external factors operating at different scales that affect the long-term viability of those land use types. This allows to link results obtained at the plot and farm levels, such as amount of C sequestered and profitability of different C sequestering livestock systems, with events and processes occurring at national and international levels, such as national payment schemes for environmental services or international negotiations for the implementation of the Clean Development Mechanism (CDM). This approach becomes relevant at the moment of translating the results of financial feasibility of different C sequestering livestock systems into policy initiatives and recommendations. In this sense, policy initiatives consider not only the restrictions and opportunities encountered at the farm for implementing C sequestering livestock systems, but also factors at the macro level that directly influence the implementation of those land use types at the farm level.

Stages of socio-economic research

To achieve the objectives listed above, the information was analysed in three stages (Table 1). Stage 1 corresponded to the socio-economic description of the farms participating in the project, and to the characterisation of conventional livestock production systems *vis-à-vis* C sequestering livestock systems regarding production levels, production costs and management conditions. This stage was mainly descriptive and static, and was based on information coming from farm surveys.

Table 1. Flow of the information in the socio-economic research.

Stage 1	Stage 2	Stage 3
Objectives		
characterisation of farm bio-physical and socio-economic conditions identification of production levels, operating costs and revenues of different livestock management systems	explore financial feasibility of investing in C sequestering livestock systems identify limitations at farm level to implement C sequestering livestock systems	development of policy guidelines to promote implementation of C sequestering livestock systems
Instruments		
farm survey and farm registers information form secondary sources	information from Stage 1 modeling incorporating risk and uncertainty	results from financial models Stage 2 information form farm's external context
Expected results		
description of bio-physical and socio-economic conditions of farms estimated parameters of production and sales for different livestock management systems	benefit-cost analysis of investing in C sequestering livestock systems	policy recommendations regarding adoption of C sequestering livestock systems

Stage 2 consisted of modeling the financial feasibility of investing in C sequestering livestock systems. Once conventional livestock production systems and C sequestering livestock systems were characterised and their production costs and sales were estimated, models were developed to explore the financial feasibility of investing in C sequestering livestock systems under different scenarios considering the presence or absence of payments for C sequestration. This stage involved dynamic modeling and integrating results from the socio-economic research with results from the bio-physical research on the capabilities of different land use systems to sequester C.

Stage 3 translated the results of the bio-physical and socio-economic research into policy initiatives and recommendations to promote the implementation of land use systems with capacity to sequester C. This phase was based on the results generated by the models on the financial feasibility of C sequestering livestock systems, and from their interpretation in the context provided by the application of the conceptual framework described above.

Farmers' participation

Farmers participating in the project in the Colombian Andean Hillsides included:
1. Those whose farms were part of an existing farm research network operated by CIPAV, which included farms that had incorporated sustainable management practices in recent years.
2. Those whose farms were representative in terms of size, ecological conditions, and land uses for the particular ecosystem, which represented farms using the traditional land use practices in the region.

Farmers participating in the Colombian Amazon region were:
1. Part of a research network being implemented by the University of Amazonia, and were in the process of implementing sustainable land use management systems on their farms.
2. A large commercial farm that was willing to provide financial information on its performance.

In the semi-humid forest of Costa Rica, farmers participating in the project were part of CATIE's research network, and were in the process of implementing sustainable livestock management practices.

Data gathering

Data needed for the socio-economic analysis were obtained from a farm survey, registers of farm production and activities and secondary sources, which was carried out between October 2002 and March 2003. Farm registers were updated once every three months between June 2003 and June 2006. The objective of the survey was to provide a detailed characterisation of the bio-physical and socio-economic conditions on the farms. The survey included questions related to the following categories of variables:
- general bio-physical characteristics of the production unit;
- general characteristics of the farmer and his/her family;
- indicators of farmer's well being;
- livestock and pasture management practices;
- availability and use of labour;
- production levels;
- production costs;
- access to credit;
- unforeseen events.

In the survey 48 farms were included: 19 in the Andean Hillsides, 20 in the Colombian Amazon region and 9 in the semi-humid tropical forest of Costa Rica. After the completion of the survey, a system of registers to record production levels, operating costs and farm activities was implemented for a small number of farms at each site, who were willing to provide information. In the eroded Andean Hillsides of Colombia farm registers were implemented at 7 farms, at 9 farms each in the humid tropical forest of the Colombian Amazon region and in the semi-humid tropical forest of Costa Rica. The objectives of implementing farm registers were to obtain detailed information on financial performance and estimates on production levels and operating costs for different land use systems. These were used as inputs to model the financial feasibility of investing in C sequestering livestock systems. The variables recorded were production levels and sales, production costs (inputs and labour); and unforeseen events.

Information from secondary sources included external factors, such as:
1. credit policies for the agricultural sector;
2. environmental policies and regulations;
3. tendencies on input and livestock prices;
4. land taxes;

5. tendencies on regional land use changes; and
6. incentives for C sequestration.

The objective of recording this information was to identify the factors that may act either as barriers or opportunities for implementing land use systems with capacity to sequester C.

Data analysis

Farm production and financial indicators

Data obtained from the survey and the farm registers were used to obtain production and financial indicators of the farms. Production indicators included volumes of products (kg/ha/yr) for milk, beef and rubber, among others. Financial indicators were computed following Wadsworth (1997) and included gross production (GP = sales + household consumption + changes in animal inventories) and gross margin (GM = GP – operating costs). These indicators were computed for each of the production systems on each farm and for each farm as a whole.

Simulation of investment scenarios

A benefit cost-analysis to evaluate the financial viability of investing in C sequestering livestock systems was conducted (Table 2). Models were developed following the methodology proposed by Brown (1997) and Gittinger (1982) to assess investments in the agricultural sector. Costs and benefits of the different land use types listed in Table 2 were estimated to assess the financial convenience of their implementation using the Net Present Value criterion (NPV). All models were based on the assumption that the relevant land use alternative was poorly managed natural pastures, the most common land use system present in the project area. Accordingly, models were developed taking poorly managed natural pasture as the starting point for the comparison.

A profit maximising farmer would switch to land use systems with capacity to sequester C if the NPV for the C sequestering livestock systems alternative (NPV_A) was higher than the one for the current land use system (NPV_C), such that $NPV_A > NPV_C$. In other words, the incremental NPV for the investment must be positive ($NPV_A - NPV_C > 0$). The incremental NPV for each of the

Table 2. Financial models developed within each sub-ecosystem.		
	Direction of the investment	
Site	From (current practice)	To (C sequestering livestock systems)
Andean Hillsides, Colombia	Natural pasture, poorly managed	Association of introduced pasture + leguminous + dispersed shade trees
Andean Hillsides, Colombia	Natural pasture, poorly managed	Introduced pasture + leguminous + dispersed shade trees + forage bank
Semi-humid forest, Costa Rica	Natural pasture, poorly managed	Introduced pasture + dispersed shade trees
Semi-humid forest, Costa Rica	Natural pastures + concentrates	Natural pastures + forage bank
Amazon Region, Colombia	Natural pasture, poorly managed	Association of introduced pasture + dispersed shade trees
Amazon Region, Colombia	Natural pasture, poorly managed	Association of introduced pasture + legumines + dispersed shade trees + forage bank

possible investments options was determined by computing the cash flow for each year of the investment's economic life. The generic notation for the models was as follow:

$$NPV = (-Ia) + \sum_{t=1}^{n} \frac{Ba - Ca}{(1 + R)^t} - \sum_{t=1}^{n} \frac{Bc - Cc}{(1 + R)^t}$$

where:

NPV = net present value of the investment, in dollars;

Ia = initial investment in the C sequestering livestock systems, in dollars;

Ba = gross benefits from the C sequestering livestock systems, in dollars;

Ca = operating costs associated to the C sequestering livestock systems, in dollars;

Bc = gross benefits from current practices, in dollars;

Cc = operating costs associated to current practices, in dollars;

R = real discounting rate, in percentage.

For each one of the models, the following steps were carried out:
1. parameters of production and sale for the current land use system and for the C sequestering livestock systems were estimated;
2. production costs and sales of each land use system were estimated;
3. cash flows over a 10 year period were calculated;
4. risk for production and price variables was incorporated, using the Monte Carlo approach;
5. expected NPV were estimated,taking into account the situation under C sequestering livestock systems management versus the situation under conventional livestock management in order to obtain the incremental net benefits due to adopting C sequestering livestock systems.

The final results for the Monte Carlo simulations were sets of 3000 NPVs for each model. Results were presented as frequency distributions of a possible range of incremental NPVs.

The basic data for milk and beef production, production costs, investment costs and sale prices were derived from the farm registers. All data were pertinent to the agricultural year 2004/2005. Livestock revenues were derived from sales of milk and meat products. Costs included:
1. operating expenditures of livestock;
2. operating costs of pasture and forage bank management; and
3. establishment costs for the corresponding C sequestering livestock systems (the investment alternative).

For the model incorporating forage banks corresponding to the semi-humid forest of Costa Rica, it was assumed that no change in productivity occurred when switching from supplementing with concentrates to supplementing with biomass from the forage banks. Accordingly, NPV was calculated for the savings incurred by replacing concentrates (an external input to the farm) by biomass produced by the forage bank. Therefore, the model compared establishment and operating costs of the forage bank vs. the cost of providing concentrates.

Data were used to create farm budgets, which were later utilised as a base for developing the models. All prices (production costs and sales) were farm-gate prices per ha. Prices were expressed in US dollars without adjustment for inflation. C data considered in the models included C accumulated in the soil plus C present in aerial biomass. The amounts of C sequestered incorporated in the models for the different C sequestering livestock systems were expressed as incremental t C/

ha/yr with respect to the current land use type (which represented the baseline C stock at the start of the project). Although the C sequestering livestock systems analysed here are not eligible under the current provisions of the CDM, the modeled payments for C were applied along a five-year period to be consistent with the guidelines for temporary Certified Emission Reductions established for the first commitment period of the CDM [see current guidelines for CER, Ninth Conference of the Parties (UNFCCC COP 9, 2003)].

Chapter 7. Socio-economic results

B.L. Ramírez, P. Cuéllar, J.A. Gobbi and J. Muñoz

Amazonia

Farm characterization

Farms participating in the study were classified in two groups (I and II):

- Type I small to medium size farms (average 66 ha) operating on family labour had a diversified array of production systems besides dual-purpose livestock with stocking rates between 0.3 and 2.8 AU/ha and milk production between 1.7 and 4.5 L/cow/day for the market and for household consumption. Production of fruit, chickens, fish, pigs, eggs and cheese for household consumption was a strategy of food security. Household consumption of farm produce ranged between 4-16% of the gross production. However, two farms devoted about half of their production (49% and 53%, respectively) to household consumption indicating that their economy was mainly based on subsistence. Type I farms represent about 75% of the farms in the Amazon region (Ramírez, 2004). About half the of the farm consisted of unimproved pastures (UP) consisting of *Paspalum, Axonopus, Homolepis* and *Calopogonium* spp. Improved pastures with *B. decumbens, B. brizantha* and *B. humidícola* covered about 25% of the farm and 18% of the farm was remnant forest. These farms also had forage banks, house-gardens and small areas of agroforestry of rubber and fruits. Animals received a daily supplementation of protein and energy from a 1 ha forage bank composed sugar cane (*Saccharum officinarum*), *Trichantera gigantea, Tithonia diversifolia, Clitoria farchildiana Cratylia argentea, Erythrina fusca, Gliricidia sepium* and *Morera* spp. However, forage banks are not common in the area. Traditionally, cattle farmers use only pastures to feed their animals and do not provide extra supplementation with concentrates.

- Type II farms were devoted entirely to extensive dual-purpose livestock production, operated with hired labour and market oriented production. There average size was 800 ha, of which 43% was forest, 35% improved pastures and 22% unimproved pastures. Average stocking rate was 0.9 AU/ha, varying between 1.1-1.3 in the area with improved pastures. Average milk production was 5.1 L/cow/day, which was higher than the average daily production of 3.2-3.8 L/cow/day reported for the region by Velásquez *et al.* (1999). The higher milk production can be attributed mainly to genetic selection of the animals and to adequate sanitary and nutritional management, which was not generally applied in the region.

Eighty-eight percent of the farmers lived on their farms and had secure property titles to their land (Muñoz, 2004). Seventy-seven percent of the farm houses were built with bricks and 86% had electric power. However, only 36% and 28% of the farms had water reticulation and plumbing, respectively.

Type I farm families consisted on average of 4.6 people, 62% of them males. Seventy-five percent of the family's members were between 11 and 49 years old. The average school attendance of women was 3.5 and of men 5 years.

Farm income indicators

Type I farms with their array of production systems, presented different structures in their production costs and exhibited a wide range in net income. Most of the farm costs (35 to 88%) was related to livestock production. Gross margins from livestock activities varied between US$ 10 and US$ 132/ha. Gross margins from rubber production ranging between US$ 104/ha and US$ 424/ha were higher than those obtained from livestock production. Farms with lower net revenues from livestock production had a tendency to use more temporary hired labour and had a higher proportion of native pastures, which were generally overgrazed. In contrast, farms showing higher revenues from livestock production tended to have a large proportion of improved pastures of *Brachiaria* spp. associated with the legume *Desmodium heterocarpon*, subsp. *ovalifolium*, which produces more forage of better quality than unimproved pastures.

On Type II farms, operating costs were associated entirely with livestock production, with labour accounting for the largest proportion (59%) of the farm's operating costs. Mean gross margin for livestock production in the Type II farm was US$ 129/ha.

Investment models

To explore the feasibility of implementing land use management systems with capacity to sequester C (LUSC), we calculated the Net Present Value (NPV) of investing in two LUSCs:
1. Improved grass-legume pasture of *Brachiaria* spp. with *A. pintoi* (Model IP).
2. Improved grass-legume pasture + forage bank (Model IP + FB).

For both models, the starting point for the comparison was poorly managed unimproved pastures (UP). The original pastures had some 20 dispersed native trees in the paddock, which were retained when the alternative LUSCs were improved. The production system modeled was dual-purpose livestock. For discounting, a real interest rate of 8% was applied (Banco de la República de Colombia, 2006). The duration of the investments was assumed as 10 years. This period corresponds to the average optimal useful life of an improved monoculture pasture of *Brachiaria* spp. The models assumed no capital limitations for making the investment in the alternative LUSCs. They also assumed that there was no decrease in productivity in the poorly managed unimproved pasture over time. All costs and benefits were calculated in US dollars (exchange rate: 2300 Colombian pesos = 1 US dollar), per ha and were pertinent to the agricultural year 2005/2006. Yearly revenues and costs per ha were based on data from the farm registers (Chapter 6).

Using the Monte Carlo simulation, the uncertainty of variations in milk and meat prices was incorporated into the models. Prices of these products show short-term fluctuations driven by factors such as weather conditions, consumption patterns, and economic conditions. Based on historical trends, it was estimated that the price of milk may deviate ± 15% from the current price of US$ 0.21/liter. Likewise, the price of meat may deviate ± 20% from the current average prices for calves (US$ 1.0/kg) and for discarded cows (US$ 0.80/kg). A lognormal probability distribution was used to model the price changes.

Description of models

Extra costs for establishment of a FB were US$ 81.74/ha (19%) (Table 1).

The stocking rate of the UP management system was assumed to be 0.6 AU/ha and cows were assumed to produce 2.0 liters of milk/day with a lactation period of 200 days. Estimated birth rate was 50%. Male calves were sold at weaning (at about 150 kg), and female calves were left in the herd to replace old cows. The rate of cow replacement was 20%/yr. Cattle management involved the provision of common and mineralised salt, vaccines, parasite control and concentrate supplementation of the animals. Management of the pasture involved a manual weeding during December and January. There was no application of fertilisers. Management costs for animals and pasture in the UP, IP and IP+FB systems are presented in Table 2.

Table 1. Establishment costs for the improved pasture association of *Brachiaria* spp. with *A. pintoi* (IP) and for the same association plus forage bank (IP+FB). Amazon region, Colombia (in US$/ha).

	Land use system	
Activity	IP	IP+FB[a]
Soil preparation	130.43	135.65
Inputs and materials	260.87	326.96
Labour	43.48	53.91
Total	434.78	516.52

[a]Referred to 400 m^2 of forage bank.

The extra management cost by changing from UP to IP was $ 20.51/ha and adding a forage bank increased it by another $ 259.14/ha.

Animals were introduced six months after establishment. The greater quantity and better quality of feed provided by the IP and IP+FB allowed increasing the stocking rate to 1.1 AU/ha. Milk production increased to 5.1 liters/cow/day in year 3 and the lactation period was extended to 220 days. Estimated birth rate was 75%. As was the case for UP, male calves were sold at weaning, but at a higher liveweight (180 kg). Cattle management of IP was similar to that of UP. Management of IP involved annual reseeding of an estimated 5% of the pasture each year to recover areas degraded by cattle walking.

The FB was productive one year after establishment. Starting at year 1, stocking rate could be increased to 2 AU/ha. Production parameters of the FB were similar to those of the IP situation. Management of the FB involved an annual manual weeding (conducted between August and November) and the application of organic fertiliser, which was produced on farm, three times a year (March, July and October). Management of cattle was similar to that of the IP system, except that animals were supplemented daily with cut and chopped material of the forage bank. Management of the pasture was similar to the described for the IP system. Management costs of Up, IP and IP +FB systems are presented in Table 2.

Table 2. Annual management costs according to land use system (US$/ha).

	Land use system[a]		
Activity	UP	IP	IP+FB
Animal management			
Mineral salt ($3.85/AU)[b]	2.31	4.23	7.70
Common salt ($3.35/AU)[c]	2.01	3.69	6.70
Vaccines ($ 3.35/AU)[d]	0.87	1.59	2.89
Parasite control ($ 6.30/AU)	3.80	6.96	12.66
Animal husbandry (4 days/AU/yr)[e]	20.87	38.26	69.57
Pasture management			
Manual weeding	26.10	17.39	17.39
Reseeding	–	4.35	4.35
Forage bank management [f]			
Application of organic fertiliser	–	–	27.39
Cut-and-carry (6 days/AU/yr)	–	–	104.35
Total	55.96	76.47	235.61

[a] UP = unimproved pasture, IP = improved pasture of *Brachiaria* spp. with *A. pintoi*, IP+BF = improved pasture of *Brachiaria* spp. with *A. pintoi* + forage bank. Labour cost = US$ 8.70/ day.
[b] Daily provision of 10 g/animal/day.
[c] Daily provision of 50 g/animal/day.
[d] Vaccines against foot-and-mouth disease, rabies (two applications/yr).
[e] Includes tasks associated with herding and milking.
[f] 400 m² of forage bank.

Model results

Under the model assumptions, investing in the replacement of poorly managed UP by IP and IP+FB was financially profitable and showed positive NPV in 100% of the trials (Table 3 and Figure 1).

When a payment for C sequestration would be applied, the expected NPV would increase 19.3% over the scenario without payment for C and they were positive in 100% of the trials. The internal rate of return (IRR) for the investment, without

Table 3. Financial analysis of investing in land use management systems with capacity to sequester C.

Land use management system [a] NP	NPV Mean (US$/ha)[b]	S.D. (US$/ha)
IP (without payment for C)	199.9	33.5
IP (with payment for C)	238.4	34.0
IP + FB (without payment for C)	353.8	65.2
IP + FB (without payment for C)	395.1	64.1

[a] IP = improved pasture association of *Brachiaria* spp. with *A. pintoi*, IP+BF = improved pasture association of *Brachiaria* spp. with *A. pintoi* + forage bank.
[b] These values represent the incremental gains over the poorly managed native pasture system.

applying a payment for C, was 20.7%. IRR increased to 22.1% when a payment for C was applied. Under both scenarios, the payback period for the investment was 4 years.

Under the model assumptions, investing in a mixed FB in association with the introduction of IP is risk free, showing positive NPV in all the trials (Table 3). When a payment for C sequestration would be applied, the expected NPV would increase 11.7% over the scenario without payment. The IRR for the investment, without applying a payment for C, would be 23.3%. IRR would increase to 24.6% with payment for C. Under both scenarios, the payback period for the investment is three and a half years.

Policy implications

The implementation of IP and IP+ FB to replace poorly managed UP in already deforested pasturelands is a financially viable option. However, the strategy must be differentiated according to farm type. The incorporation of cut-and-carry FBs is feasible in small and medium size farms (Type I) but impractical in large farms (Type II) due to their high demand of labour and the fact that large farms have the resources to intensify production by other more scale-efficient means.

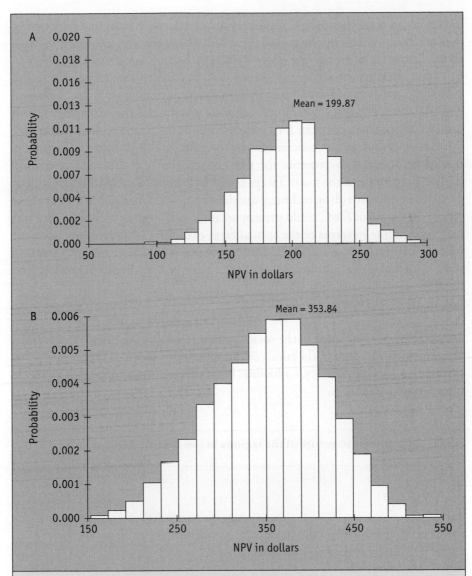

Figure 1. (A) Probability distribution for incremental net present values (NPV) for investing in a 1-ha of improved pasture of *Brachiaria* spp. in association with *Arachis pintoi* from poorly managed native pasture. Amazon region, Colombia. (B) Probability distribution for incremental net present values (NPV) for investing in 0.04 ha of fodder bank plus 1-ha of improved pasture of *Brachiaria* spp. in association with *Arachis pintoi* from poorly managed native pasture. Amazon region, Colombia.

The effect of a payment for C sequestration is relatively marginal in terms of making the investments financially feasible. However, the existence in large farms (Type II) of important areas with remnant of forest indicates that in this type of farms the strategy, in terms of C payments, should be oriented to avoid deforestation and to promote incorporation of trees in pastures. Although Type II farmers posses the capital to make the initial investments without external support via payment for C sequestration, the need technical assistance to manage grass-legume associations. In the case of Type I farms, the strategy should be oriented to promote the implementation of improved grasses and legume pastured and the use of forage banks by means of subsidising the initial investment and providing technical assistance for their management. For these farms, the payment for C could present an additional source of income, but not a significant one in terms of the annual farm cash flow. However, any payment scheme involving these small and medium farms would have high transaction costs, which may preclude making it effective as a policy intervention tool.

Andean Hillsides, Colombia

Seven farms were included in the socio-economic study of the Andean Hillside. These farms are located in the rural communities of El Dovio and Dagua (1,450-1,800 m.a.s.l.), Cauca Valley, Colombia. Basic infrastructure of both regions are summarised in Table 4.

Table 4. Basic infrastructure of the regions in the Andean Hillsides.

	El Dovio	Dagua
Primary school	Nearby	Nearby
Secondary school	Far	Nearby
Hospital	Far	Nearby
Market	Far	Nearby
Public transport	Twice a week	Daily
Aqueduct	Community service	Community service
Electrical energy supply	Private company	Private company
Communication	Public Telephone and Cell Phones	Public Telephone and Cell Phones

The most relevant changes in land use and in farm production systems in the communities are shown in Tables 5 and 6.

Farm types studied in the project and their characteristics

Two types of farm were studied during the time of project: (1) improved farms, where the farmer had implemented sustainable production systems, and (2) traditional farms, which maintained the traditional way of farming of the region.

Table 5. Land use changes in El Dovio. Modified after Espinel (1994) and Rosero (2000).

Before 1950s	Native forest	
1950 to 1970	Cash crops	Coffee, plantain, tamarillo (*Cyphomandra betacea*)
	Subsistence crops	Maize, beans and some vegetables
	Native pastures	*Axonopus scoparius*, *A. micay* and *Paspalum notatum*.
		Adapted forms of *Melinis minutiflora*
Mid 80s	Introduction of pastures	*B. decumbens* and *C. plectostachyus*
End 80s	Introduction of forage species	*Saccharum officinarum*; *Trichanthera gigantea* and *Erythrina edulis*
Present	Subsistence crops	White carrot, manioc, maize, plantain, beans and vegetables
	Farming systems	Animal production and feed: diversified or extensive livestock production; mainly dual-purpose cattle, pigs and poultry
		Plants used for animal feed: *Trichanthera gigantea, Tithonia diversifolia, Xanthosoma saggitifolium, Morus alba, Erythrina edulis, Boehmeria nivea, Azolla anabaena, Lemna minor*
		Agropastoral systems: Introduction of woody trees (*Montanoa quadrangularis; Cordia alliodora*) and fruit trees (guava *Psidium guajaba*)

Table 6. Land use changes in Dagua.		
Before 1970s	Native forest	
Mid 1970s	Cash crops	Coffee, pineapple, flowers (*Anthurium andreanum*)
	Subsistence crops	Plantain, maize, beans
	Improved pastures	*B. decumbens* and *C. nlemfuensis*
End 1980s	Introduced forage species	*Trichanthera gigantea* and *E. edulis*
Present	Cash crops	Pineapple in extend areas and coffee in small areas
	Subsistence crops	Plantain and beans
	Farming systems	Animal production systems: Mainly dual-purpose cattle, pigs and poultry for meat production.
		Animal feed: Foliage of *T. gigantea, T. diversifolia, X. saggitifolium, M. alba, Erythrina fusca, E. edulis, E. poeppigiana, Gliricidia sepium, Urera caracasana, Boehmeria nivea, Smallanthus riparius*

The most relevant characteristics of the improved farms were:

- Highly diversified and commercially orientated animal production with more than two animal species. Additionally, some of the farmers were processing milk and meat. Diversified production favours income stability.
- Use of mixed forage banks (protein and energy crops) to improve the farm's forage resources for cattle, pigs and free-range poultry. Generally, the farmers establish forage trees, shrubs, forage grasses and sugar cane, cultivated at high density and in different associations to make maximum use of the available area. Forage banks were cut and the material brought to the stables (cut-and-carry systems).
- Implementation of silvopastoral systems, such as pastures with dispersed trees planted by the farmer or developed by natural regeneration, live fences with diverse trees or shrubs for animal feed and/or wood, especially *Montanoa quadrangularis* and *Cordia alliodora*.
- Recycling of organic matter and waste waters using tubular plastic biodigestors. Effluent water of the biodigestors is used in fertilisation and irrigation. Resulting biogas substitutes firewood and reduces the amount of commercial energy sources for cooking like propane gas and electricity. Some

of the farmers use biogas in simple but functional heating systems for piglets and chicks.
- Watersheds conservation and protection.
- Low to medium access to capital for investments and for the management of the farms.

The traditional farms of the region are characterised by:
- Small scale monoculture cropping, mainly pineapple (*Ananas comusus*), runner beans (*Phaseolus coccineus*) and in some areas coffee (*Coffea arabica*), both for sale and family consumption.
- Extensive cattle grazing on poorly managed pastures mainly composed by unimproved grasses.
- Little or no use of the organic matter produced by the farming system. Generally, chemical fertilisers are bought in the market and used only for cash crop production.
- The availability of capital for investment or farm management is low.

The distribution of farm area according to the land use is shown in Table 7. Farms in the survey ranged from 1.8 (1 farm) to 20 (1 farm) ha, with a mean of 5.5 ha. Average farm size in the municipality of Dagua was 5 ha, and in the municipality of El Dovio farms in the survey were very small, between 0.3 and 5.0 ha, and medium sized between 10 and 20 ha.

Table 7. Distribution of the farm area according to land use.

Type of farm	Forest	Forage banks	Improved pastures	Unimproved pastures	Crops
Improved farm	30.0%	18.0%	13.0%	13.0%	26%
Traditional farm	20.2%	0.3%	8.3%	32.3%	39%

Farming systems in the Andean Hillsides

Cattle raising was the common land use on the farms studied. On traditional farms, management was extensive with cattle grazing unimproved and poorly managed pastures of *Hyparrenia rufa*, *Melinis minutiflora* and *Paspalum notatum.* Use of improved grasses, mainly African star grass (*C. nlemfuensis*) was beginning in small areas. The farmers also keep milking cows for home milk consumption, but also to fatten calves by reserving one or two quarters of the uddder for calves.

Cattle production systems on improved farms were both dual-purpose and beef production. Milking cows on improved farms were kept in confinement in a production system of restrained suckling for calves: in the first three months, one quarter of the udder is left for the calf, then it gets residual milk. In dual-purpose systems, biomass of mixed forage banks are commonly used by farmers to feed their cows (cut-and-carry). Young cattle graze both on unimproved and improved grasses like *C. nlemfuensis* and *B. decumbes*, associated with legumes such as *A. pintoi.* Live fences are widely established in all farms.

In beef production systems, beef cattle are reared on pastures composed of *Brachiaria* spp. associated with *A. pintoi*, or on native pastures with trees, mainly *Psidium guajava.* Nearly all the forage from mixed forage banks is destined for pigs and poultry feed.

Pig production is common on improved farms, producing fattening pigs between 85 and 90 kg liveweight. Pigs are fed forage from mixed forage banks and sugar cane plus restricted amounts of concentrate feed (cereals and oil meal).

With regard to poultry, improved farms rear broilers and hens for meat and egg production. Broiler feed system is based on 100% commercial concentrates. Hens are held in free-range systems, roaming the cropping areas. Broiler meat is sold directly by the farmers. Traditional farms keep very commonly exclusively free-range poultry: hens, ducks and turkeys. Meat and eggs are both for sale and family consumption.

Socio-economic base-line of the studied farms

Important characteristics of the socio-economic status of farms were colleted at the beginning of the project (Table 8).

Farm productivity and labour force

The productivity of the farming systems (ha/year) and the indicators of livestock productivity are presented in Tables 9 and 10.

Table 8. Socio-economic indicators of the farmers in the Andean Hillsides sub-ecosystem.

Indicator	Farms	
	Improved (N = 7)	Traditional (N =12)
Adult literacy (%)	79	76
Average years of schooling	8	6
Farmers, who are completely dedicated to their farms (%)	40	32
Family group size (persons)	1–7	4–5
Land tenure	> 15 years	

Table 9. Milk production of the farming systems (L/ha/year).

Indicators	Farm type and mean area		
	Improved Farm I (1.8 ha)	Improved Farm II (5.0 ha)	Traditional Farm (18.0 ha)
Dual-purpose cattle	4,421.4		
Beef cattle		501.0	204.6
Destination of milk production			
Processing of dairy products (mainly cheese)	42.0%		
Market	55.6%	15%	21%
Family consumption	2.4%	85%	79%
Livestock and other products (for market)			
Cattle, tons of live-weight	1.3	0.77	0.56
Pigs, tons of live-weight	0.93	0.18	-
Eggs, total farm production (units)		1,560	3,123.0
% for family consumption		35%	25%
Coffee, tons		-	300.0

Table 10. Parameters of cattle production.

Indicators	Farm type		
	Improved Farm I (1.8 ha)	Improved Farm II (5.0 ha)	Traditional Farm (18.0 ha)
Cattle extension area, ha	1.5	3.5	5.7
Milking cows farm/day	3	1	1.1
Litres of milk/cow/day	6.1	4.5	2.8
Calving interval, months	15.4	14.1	22.6
Stocking rate AU/ha	4.6	1.5	0.88
Number of sold livestock/ha/yr			
Weaned calves	1.0	0.28	0.35
Beef cattle		2.3	0.61
Culled bulls and cows			0.87

Data are shown for the different farm types. According to the cattle production system, the improved farms are subdivided into two groups: Improved Farms (I) with dual-purpose cattle production systems and Improved Farms (II) with beef cattle production:

1. Improved Farm I:
   a. Dual-purpose cattle production.
   b. Milking cows kept mainly confined.
   c. Cut-and-carry systems from mixed forage banks for cattle and pig diet.
   d. Limited use of external feed inputs (concentrate feed, salt and minerals).
   e. Young cattle grazed pastures with low density of trees, and live fences.
2. Improved Farm II
   a. Dual-purpose cattle production, focused on beef cattle.
   b. Cattle grazing rotationally paddocks with low density of trees (natural regeneration).
   c. Feed of mixed forage banks for pigs and poultry.
   d. Limited use of concentrate feed for cattle and pigs.
3. Traditional Farm
   a. Calf rearing for beef production.
   b. Extensive cattle production on unimproved grasslands.
   c. Minimal or no use of feed from mixed forage banks.

d. Scarce use of concentrate feed for cattle. Maize and cooncentrate feed were bought for free-range poultry.

The use of veterinary drugs is common by all farm types, even though they do not fulfil the sanitary programs of the zone.

Half of the improved farms have an external labour force. All of the improved farms and 66% of the traditional ones hire workers for specific tasks. All farm types use family labour. The kind of labour force and the number of day wages on the surveyed farms are given in Table 11.

In the Improved Farm I, 85% of the work is employed for dual-purpose cattle production, the remaining 15% for the processsking of dairy products and pig production. About 65% of the work in cattle production is employed in the maintenance of silvopastoral systems and for the cut-and-carry of forage.

Table 11. Kind of labour force and number of day-wages per year.

Labour force	Farm type		
	Improved Farm I (1.8 ha)	Improved Farm II (5.0 ha)	Traditional Farm (18.0 ha)
Permanent	62%		
Temporary	13%		16%
Family	25%	100%	84%
Day-wages per year	504	229	169

Economic analysis

Economic information has been considered in US$/ha/year to compare the different farms. Economic indicators refer to gross production, gross margin and net income. Management and administration income (MAI) indicates if the cost of family labour is covered by the farm income (MAI >0). The results of economic analysis, made for the different farm types is shown in Table 12.

Table 12. Farm income, US$/ha/yra.

Indicators	Farm type		
	Improved Farm I (1.8 ha)	Improved Farm II (5.0 ha)	Traditional Farm (18.0 ha)
Gross production	2,937	570	193
Gross margin	695	247	146
Net income	583	231	72
MAI	> 0	> 0	< 0

aValues are expressed in US dollars using the average exchange rate of Col.$ 2,500 pesos to the dollar.

On all farms, families get additional income from external sources, like jubilation subsidies, temporally employment in the construction sector, manual work or community work.

Financial models for the different types of land use

Two models of an *ex-ante* analysis of costs and benefits will be discussed to evaluate the monetary profit of investing in different kinds of land use. Depending on the different production systems, the proposed economic models are:
- Model 1. *'IP + DT'*
  Improved pastures (IP) plus legumes (*Brachiaria* spp. + *A. pintoi*) with dispersed tree cover (DT) (50 trees/ha. Change of land use 100%.
- Model 2. *'IP + DT + MFB'*
  Brachiaria spp. and *A. pintoi* + dispersed tree cover and mixed forage banks (MFB). Establishment of the forage bank on 50% of the area.

Original land use of both models was a poorly managed native pasture. Models were run on dual-purpose cattle production. The assumptions used in economical analysis were:
- Source of Data: Productivity of farms in the survey (2003-2006).
- Constant prices, average exchange rate of Col.$ = 2,500 pesos to the US dollar.

- Real Discount Rate (DR) equivalent to 9%.
- Investment time horizon: 10 years.
- Financial indices: Internal Rate of Return (IRR) and Incremental Net Present Value (NPV).

The results are shown in terms of 1 ha. Cash flow was calculated on Net Present Value (NPV).

Description of the models

Production systems on poorly managed unimproved pasture

The extensive cattle production on poorly managed unimproved pastures supported 0.88 AU/ha. Average milk production was 2.8 L/cow/day, with an estimated lactation length of 200 days. Calving interval was 22.6 month, with a parturition rate of 53%. Milking system was by hand with calf close by. Weaned calf live-weight was about 130 kg.

Cattle were sold as breeding heifers and beef cattle, with a cow culling rate of about 17%. Additional to forage, there was a limited offer of minerals or common salt mixed with molasses of sugar cane. Another common feature of the system was the use of veterinary drugs. Fertilisation and irrigation of pastures was scarce, twice a year hired labour was employed for weed control. Family labour predominated.

Production systems with improved pastures, legumes and dispersed trees 'IP + DT'

Cattle farming systems on improved pasture, such as *Brachiaria* spp. or African star grass (*C. nlemfuensis*). with *A. pintoi*, had milk production of 4.5 L/day. The stocking rate fluctuated between 1.5 and 1.7 AU/ha, calving interval was reduced to 14,1 month, giving lactation periods of about 240 days and a parturition rate of 85%. Weed control in pastures was manual, the fertilisation consisted of organic matter and the effluent of the tubular plastic biodigestor was used for irrigation.

Cattle feed included multi-nutrient blocks, salt and minerals. Additionally, sugar cane tops and a limited amount of forage from the forage banks were given. The calves were reared on the farm and sold as breeding heifers or beef cattle. The milking system wss by hand with the calf close by.

Production systems with improved pastures, legumes and dispersed trees 'IP + DT+ MFB'

Milking cows were kept in confinement, calves, dry cows and heifers were put to pasture. The principal characteristic of this system was cut-and-carry of forage from mixed forage banks, which included trees, shrubs, forage grasses and sugar cane. The stocking rate was about 4.6 AU/ha and the milk production 6.1 litres/cow/day. The calving interval was about 15.4 months and parturition rate 78%. Milking system was twice a day by hand, with the calf close by. The weaned male calves are sold at an approximate live-weight of 170 kg. Additional cattle feed consisted in a limited offer of concentrates for milking cows (57 g per litre of milk and mineral salts), calves received concentrate feed only during the first three months.

Establishment costs of silvopastoral systems

Establishment costs of improved pastures in association with legumes, dispersed trees and mixed forage banks are shown in Table 13. Management of forage banks included manuring with organic matter (fresh manure from cattle, deep litter from poultry), irrigation with the biodigestor effluent and the re-sowing of forage species. The major part of labour was dedicated to these activities.

Table 13. Establishment costs of silvopastoral systems (US$/ha).

Land use	Costs US$/ha		Description
	Establishment	Maintenance [a]	
Improved pasture + legume with dispersed tree coverage	417.5	104.0	*C. nlemfuensis, Brachiaria* spp., *A. pintoi, Montanoa* spp., *Psidium guajava, Inga* spp.
Mixed forage banks	1,272.5	177.0	Trees, shrubs, herbaceous, sugar cane and *A. pintoi*

[a]During investment horizon (10 years).

Results of the models

Table 14 shows the healthy IRR and NPV with and without C sequestration payments.

Table 14. Internal rate of return IRR and incremental net present value NPV of the resources used.				
Model	C sequestration payment (/t/ha)			
	IRR %		NPV US$/ha	
	Without	With	Without	With
Model 1. 'IP + DT'	18.8	19.5	244.4	261.6
Model 2. 'IP + DT+ MFB'	21.3	21.4	1,093.2	1,104.5

Conclusions

1. Livestock productivity and economic results of Improved Farms demonstrated that the improved farming system is profitable. Productivity of improved farms allows farmers to employ permanent workers.
2. The *ex-ante* analysis of financial viability, which was used to evaluate the profitability of investments in a change of land use from traditional farming systems to improved silvopastoral systems, shows a high profitability. Investment recovery is possible over a period of two years. Although the financial model did not consider the payment of any accumulated C, the model results were economically attractive.
3. Establishing a more sustainable farming system in the Andean Hillside sub-ecosystem, that allows:
   a. to improve quality and amount of livestock feed and to diminish the dependency on external inputs;
   b. to control erosion, minimising the area of extensive grazing;
   c. to recycle organic matter and waste waters;
   leads to a wide range of benefits. These benefits include improved soil quality, conservation of water resources and an income that increases family welfare and offers employment for the local community.

4. The economical benefit of a payment for C stored with this type of land use, could be used as an incentive to strengthen the capacity of the Andean Hillsides farmer to invest in the improvement of his farming system.

Costa Rica

Characteristics of the region

The Central Pacific Region constitutes a strip between the sea and the mountains along the central portion of the Pacific coast of Costa Rica and encompasses some 3,900 km^2. The region has an estimated population of 200,000 inhabitants. The main city in the region is Puntarenas (25,000 inhabitants). The unemployment and illiteracy rates for the region are 6% and 10%, respectively (INEC, 2001). About 29% of the population in the region is considered poor, which is slightly higher than the national average (21%). Main economic activities in the region are livestock production (meat and milk) and commercial crops, such as sugar cane, rice and tropical fruits. The region possesses good infrastructure in terms of roads, it is dissected by the Panamerican Highway, several paved roads and an extensive network of secondary unpaved roads. This gives easy access to the main regional market (city of Puntarenas) and with the Central Highlands, the capital city of San José and the milk-processing plants in Monteverde and Zapotal.

According to the Agricultural Census of 2000, there were 2,729 farms engaged in livestock production in this region of Costa Rica, which is 7% of the total livestock farms in the country (MAG, 2001). The average size of livestock farms in the region is 50 ha, which is larger than the national average (35 ha). The main reasons for the larger livestock farms in the region are poor soil quality and the presence of a long dry season, which causes marked seasonality in pasture quantity and quality. The total cattle herd in the region is of 126,300 head, which is approximately 11% of the national herd, with an average of 36 head per farm. The average herd is composed of 69% females, 35% cows older than 3 years, 18% males younger than two years and 18% by calves less than one year old. About 96% (136,500 ha) of the land used for livestock production in the region consists of grasslands, mostly of unimproved species. It is common to find small areas with remnants of forest or with secondary forest regeneration on the farms, particularly on those larger than 10 ha.

The production unit consists generally of one farm, although there are farmers with more than one farm (usually two), who use the second farm as a forage reserve for the dry season. In such cases, animals are transported to the second farm during the dry season to reduce the stocking rate and allow pasture regeneration at the main farm. The most common livestock production systems in the region are dual-purpose and beef (fattening of calves); farms specialised in milk production are rare.

Dual-purpose livestock production system. Farms with this production system produce milk, which, apart from home consumption, is sold either as fluid milk or as cheese and to raise calves, which are sold at weaning. Revenues generated by these two products are generally 30% from milk and 70% from beef (calves plus discarded cows). The most common cattle breeds are crosses of Brahman with Holstein, Jersey and Brown Swiss (Pardo Suizo).

Beef livestock production systems. These are pure beef production farms that either raise calves to be sold at weaning at a liveweight of 150-180 kg (between 8-10 months of age) or to be sold when they reach a liveweight of around 300-400 kg.

Cattle management includes the provision of common and mineralised salt, vaccines, and parasite control. Cows are supplemented during the dry season (December to May) with a mix of chicken dung plus honey. Management of the unimproved pastures (UP) involves two manual weedings, one during the dry and another during the rainy season, along with the application of herbicides. There is no application of fertilisers. Management costs for animals and for the pasture in the UP system are presented in Table 15.

The dual-purpose livestock production system is common on farms up to 80 ha, with smaller farms concentrating on milk production. Beef production is common on farms larger than 40 ha and it is almost exclusive on farms larger than 100 ha. However, the region, like the whole country, is showing a tendency toward the transformation of dual-purpose farms into beef farms. This transformation responds to the significant increase in the demand for beef in the Central American region during the last few years.

Farm characterisation

Farms participating in the socio-economic study were livestock farms with dual-purpose production systems. Average farm size was 57 ha (range 15-134 ha). Most of the farms (89%) are of small and medium size (less than 85 ha), while the rest

Table 15. Annual management costs for unimproved pasture (UP) and for improved, plus trees (IP+trees) (in US$/ha).

	Land use type[a]	
	UP	IP+trees
Pasture component		
Inputs[b]	18.03	4.51
Labour[c]	61.48	24.59
Animal component		
Inputs[d]	50.47	94.63
Labour[e]	29.51	55.33
Tree component		
Labour[f]	-	6.15

[a]UP = unimproved pasture, IP+trees = improved pasture of *Brachiaria* spp. with 30 trees/ha.
[b]Herbicides (US$ 5.51/l).
[c]Weeding and herbicide application (US$ 6.15/labour day).
[d]Includes provision of mineralised and common salt, application of vaccines, supplementation with chicken dung during the dry season, and application of vermin killers (US$ 63.08/animal unit).
[e]Includes milking, supplementing and herd management (36.88/animal unit).
[f]Involves weeding during the first three years (US$ 6.15/labour day).

are larger than 100 ha. Soils are poor in structure and fertility. Soil erosion levels vary from light to severe. Options for growing crops in the region are limited by poor soil conditions and the severe dry season.

Grasslands account for 80% of the land use, of which 70% consist of unimproved grasses and the remainder of improved species (*B. brizantha* and *B. decumbens*). Small areas of herbaceous legumes (e.g. *A. pintoi*) and of forage banks (FB) are present on only two farms. Pasture subdivision is for 70% by electric fences. There is only a low density (2-5/ha) of trees dispersed in pastures.

There are remnants of natural forests, either on steep slopes or along creeks, covering between 4% and 31% of the farm area. Larger farms generally have more forest. Areas of secondary forest regeneration are not common with this land use.

Farm infrastructure consists, generally, of the family house and a corral to handle the animals and milk the cows. The average distance from the farm to the nearest town is 2 km. Estimated land values in the area range between US$ 1,025 and US$ 2,565 per hectare.

Farmer characteristics

The study involved 9 families with a total of 39 people. In the sample, 64% of the family members were between 15 and 59 year of age and 38% of were males older than 15 years. All farmers resided permanently on the farm with their families. The average family consisted of 2-7 persons. The head of the household was on average a 46 year old male. Farmers participating in the study had secure land tenure of their farms, and had owned the farm during at least 12 years. It is common for farmers to inherit the farm from their parents.

Excluding underage school children, there was only one illiterate farmer among the nine families interviewed. In general, the farmer and his family can at least read and write and have attended primary school. All children of school age attend school. The main activity of 44% of the family members older than 10 years was working on the farm for 42 hrs/week.

The farm house was generally made of bricks, combined with wood, with a zinc roof. The house had 4 to 6 rooms, running water and electricity. Electronic appliances, such as television sets, radios, stoves and fridges were commonly present. Eightyeight percent of interviewed farmers had a truck, usually more than 8 years old.

Farmers have easy access to credit. Besides the public and private sources of finance, there are different organisations in the region allowed by law to provide finance to farmers, including the Esparza Agricultural Center (EAC). Transaction costs to obtain loans are relatively low. In the last five years, three of the nine interviewed farmers obtained loans from the private sector and from the EAC. The loans were mainly to buy animals and had a repayment period of one year.

Investment models

To explore the feasibility of implementing land use management systems with capacity to sequester C (LUSC), NPV of investing in two LUSCs was calculated:

- Model IP: improved pasture composed of *Brachiaria* spp. with incorporation of 30 trees through natural regeneration, and
- Model FB: IP plus supplementation with biomass from a mixed FB of *Cratylia argentea* and sugar cane.

All costs and benefits were calculated on a hectare basis in US dollars (exchange rate: 488 Colones/US dollar), and were pertinent to the agricultural year 2005/2006. Yearly revenues and costs per ha were based on data from the farm registers (Chapter 3). The discount rate was 8%.

IP Model

For this model the starting point for the comparison was poorly managed unimproved treeless pastures (UP) of *Hyparrenia rufa*. The production system was dual-purpose and the duration of the investments was assumed to be 10 years. This period corresponds to the average optimal useful life of a monoculture pasture of *Brachiaria* spp. The model assumed no capital limitations for making the investment in the alternative LUSC.

Establishment costs for the improved, introduced pasture of *Brachiaria* spp. (IP) were US$ 161, which included soil preparation, fertiliser application and sowing. Animals were introduced six months after the establishment of the IP. The greater quantity and quality of feed of the IP increased the reproductive and productive parameters of the herd. Stocking rate was increased to 1.5 AU/ha and milk production per cow was 5.2 l/day, estimated birth rate was 75% and the inter-birth period was reduced to 16 months. Male calves were sold at weaning at a liveweight of 180 kg and female calves were kept as replacement for older cows at 15% per year. Milk production increased to 702 l/ha in year 1 and to 936 l/ha/year in year 2 and meat production increased to 114 kg/ha in year 1, to 136.8 kg/ha in year 2 and to 152 kg/ha/year afterwards.

Cattle management was similar to the UP situation. Management of the IP was similar to the UP situation, except that trees growing in the pasture by natural regeneration were not removed. Management costs of UP and IP are presented in Table 15.

Using the Monte Carlo simulation, the uncertainty of variations in milk and meat prices was incorporated into the model. Prices of these products undergo short-term fluctuations driven by factors such as weather conditions, consumption patterns and economic conditions. Based on historical trends, it was estimated that the price of milk may fluctuate ± 20% from the current price of US$ 0.20 per liter. Likewise, the price of meat may fluctuate ± 15% from the current average prices for calves of US$ 1.17/kg and for discarded cows of US$ 0.97/kg. A triangular probability distribution (minimum, most likely, maximum) was used to model the price changes.

Several studies have shown that dispersed trees (between 15% and 25%) (e.g. Restrepo *et al.*, 2004; 2004, Souza de Abreu *et al.*, 1999), could contribute to the sustainability of silvopastoral systems as a whole, producing additional income, providing shade to limit heat stress in cattle, and also helping to conserve biodiversity. Nevertheless, farmers are reluctant to incorporate trees in pastures. The concern is that shade trees may reduce pasture growth, which in turn may reduce milk and beef production. To reflect this concern, the model assumed decreases in milk and meat production of up to 10% as trees grow and the canopy begins to close. A linear decreasing function was used to express the decrease in production with the following values in percent: year 7 (-5%); year 8 (-7.5%); years 9-10 (-10%).

According to the results from the project's bio-physical component, poorly managed UP had an annual rate of C sequestration of 0.6 ton/ha, whereas IP with trees present had an annual C capture of 3.7 ton/ha. Therefore, changing from poorly managed UP to IP with trees yields an annual incremental rate for C capture of 3.1 t/ha. Based on that information, a scenario with a payment under the Temporary Certified Emission Reductions (tCER) scheme for the incremental amount of C sequestered by the introduction of the IP plus trees was modeled. The payment was applied at year 5 and year 10, the accumulated incremental amounts of C were 15.5 ton and 31.0 ton, respectively. The estimated price per ton of C for the tCER scheme in the region was US$ 2.5.

IP model results

Under the model assumptions, investing in IP was financially profitable with positive NPV in 100% of the trials (Table 16). When payment for C sequestration was applied, the expected NPV increased with 26.5% over the scenario without payment for C and they were positive in 100% of the trials (Table 16). The results also indicate that the modeled decrease of up to a 10% in the production of milk and meat due to a potential negative shade effect generated by the trees did not make the investment unprofitable in any of the modeled scenarios.

Table 16. Financial analysis of investing in introduced, improved pasture of *Brachiaria* spp. plus trees.

Land use management system[a]	Net present value (NPV)			Internal rate of return (IRR)		
	% over 0	Mean expected value (US$/ha)[b]	S.D. (US$/ha)	% over 0	Mean expected value (%)[b]	S.D. (%)
IP+trees (without payment for C)	100.0	217.4	11.4	100.0	26.4	1.12
IP+tress (with payment for C)	100.0	275.0	11.3	100.0	28.9	1.13

[a]IP+trees = introduced pasture of *Brachiaria* spp. with 30 trees/ha.
[b]These values present the incremental gains over the poorly managed native pasture system.

FB model

This model compares the costs incurred by supplementing milking cows either with chicken dung and honey or with forage produced by a mixed forage bank of *C. argentea* and sugar cane (*Saccharum officinarum*). Consequently, the model estimated the savings in implementing the FB. It was assumed that milk production remained the same under both supplementation regimes. The duration of the investments was assumed for 12 years, which corresponds to the average expected useful life of the mixed FB. The model assumed no capital limitations for making the investment in the new LUSC.

Establishment costs for the 0.9 ha FB of 0.3 ha of *Cratylia* plus 0.6 ha of sugar cane were US$ 659.76. It was estimated that 0.9 ha of a mix forage bank would produce enough biomass to supplement 20 milking cows during 120 days in the dry season (Holguín *et al.*, 2003). FBs are productive one year after their establishment. Starting in year 1, cows were supplemented daily with 10 kg of sugar cane and 5 kg of *Cratylia* per animal. To do so, the Cratylia and the cane were cut and carried from the FB to the barn, where it was chopped with a chopping machine and fed to the cows. Management of the FB included weeding every three months, and one pruning with incorporation of cut material to the soil during the rainy season. Fertiliser was not applied. Annual management costs for the forage bank are presented in Table 17.

In the current situation, milking cows are supplemented during the dry season with a daily ration of 5 kg of chicken dung and 0.3 kg of honey per animal. Current prices for these supplements are US$ 0.04/kg for chicken dung and US$ 0.16/kg for honey. The supplementation period was estimated at about 120 days

Table 17. Annual management costs for a mixed forage bank of *Cratylia* and sugar cane[a,b].

	Cost (US$)
Manual weeding	33.40
Pruning[c]	14.20
Cut-and-carry[d]	368.85
Energy[e]	86.32

[a]Labour cost/day = US$ 6.15.
[b]Forage bank = 0.3 ha of *Cratylia argentea* and 0.6 ha of sugar cane.
[c]Included incorporation of cut material into the soil.
[d]Included the daily cutting, carrying and chopping and feeding of 10 kg of cane and 5 kg of Cratylia, and during 120 days.
[e]Included the cost of energy for the chopping machine.

(mid January to mid May). Supplementation costs for 20 milking cows during 120 days was estimated at US$ 606.60.

A critical aspect regarding the cost of supplementation was the number of days that animals needed to be supplemented as this depends on the duration of the dry season. To take this aspect into account, a triangular probability distribution was used to model the minimum (90 days), the most likely (120 days) and the maximum (135 days) number of days that animals had to be supplemented with either chicken dung plus honey or with forage from the FB.

Similar to the IP model, a scenario with a payment under the tCER scheme for the incremental amount of C sequestered by the introduction of the mixed FB was modeled. The payment was applied in year 5 and in year 10, the accumulated incremental amount of C for those years were 3.0 ton and 6.0 ton, respectively. The reason for the low C sequestration rate is that most of the biomass of the FB is removed to be converted into animal production. The price per ton of C was US$ 2.5.

FB model results

Under the model assumptions, investing in a mixed FB of Cratylia and sugar cane was financially profitable and showed positive NPVs in almost all of the trials (Table 18). When a payment for C sequestration was applied, the investment was risk free with expected NPV positive in 100% of the trials (Table 18).

Policy implications

Following are the main policy implications for the socio-economic results for the sub-humid tropical forest of Costa Rica.
1. The introduction of improved grass pastures with trees and forage banks present a high potential as a strategy for sequestering C since they can be used for recuperating already deforested, degraded pasturelands.
2. Both land use types have the potential to increase production and income of the farm. The implementation of forage banks substantially increased the use of labour on the farm, which creates an additional source of labour in the area.
3. High establishment costs, in particular of forage banks, coupled with limited availability of capital of small and medium farmers may present a barrier for implementing those land use types.

Table 18. Financial analysis of investing in a mixed forage bank of *Cratylia* and sugar cane.

Land use management system[a]	Net present value (NPV)			Internal rate of return (IRR)		
	% over 0	Mean expected value (US$/ha)[b]	S.D. (US$/ha)	% over 0	Mean expected value (%)[b]	S.D. (%)
BF (without payment for C)	99.0	69.7	24.9	99.6	10.1	0.76
BF (with payment for C)	99.9	81.0	24.9	100.0	10.9	0.74

[a]FB = Mixed forage bank of *Cratylia argentea* and sugar cane.
[b]These values presen the incremental savings over supplementing 20 milking cows with chicken dung and honey during 120 days.

Instruments for policy intervention to favour implementation of improved pastures with trees and for forage banks, and thus contribute to reduce poverty, should include:
1. 'green' credits, with the loan consisting of a portion devoted to finance a component associated with production, such as forage banks, and another to conservation, such as allowing trees in pastures;
2. technical assistance in managing pastures, trees and forage banks; and
3. provision of inputs, such as tree seedlings.

The effect of a payment for C sequestration was relatively marginal in terms of making the investments financially profitable. However, the potential for the farms in the area to participate in a payment system for C sequestration is intermediate, given that transaction costs associated to payment schemes are highly dependent on farm size (the smaller the farm, the higher the transaction costs). In addition, critical issues associated to the Clean Development Mechanism such as leakage, additionality and permanence need to be considered if a payment for C is recommended. Nevertheless, the option of payments for 'avoided deforestation' (topic under discussion in the Convention on Climate Change) may offer a potential opportunity in the future.

Chapter 8. Reflections on modelling and extrapolation in tropical soil carbon sequestration

B. Van Putten and M.C. Amézquita

Introduction

In this chapter, we deal with modelling and extrapolation of C storage in tropical soils. Although our main goal is to reach recommendations and conclusions for some main research questions of the C sequestration project as described extensively elsewhere in this book, we will also discuss general aspects of modelling and extrapolation methodologies that are applicable in a much broader context.

We start by describing the C-sequestration project from a modelling point of view. This implies that, among others, abstract notations for Land Management Systems (A, B, C, ...) will be used instead of their real names, and that some minor complications are neglected that were encountered during the implementation of the experimental design (like the non-occurrence of each Land Management System (LMS) in each farm, which was easily solved by combining farms). Moreover, in the reality of the C-seq. project, the number of farms and Land Management Systems involved in the project were not necessarily equal to what is presented here, but from a modelling point of view this is irrelevant.

Along these lines, a description of the C-sequestration experimental design is as follows (see also Figure 1).

The C-sequestration project investigated two ecosystems in Colombia and two in Costa Rica. The analysis is carried out *per* ecosystem, as a specific Land Management System (LMS) may perform differently in each ecosystem, possibly resulting in different C sequestration capacity.

In our model, a total of eight farms is selected in each ecosystem, each farm containing the Land Management Systems A, B, C, ..., H under study. From now on, we will use the statistical term 'treatment' to indicate a LMS.

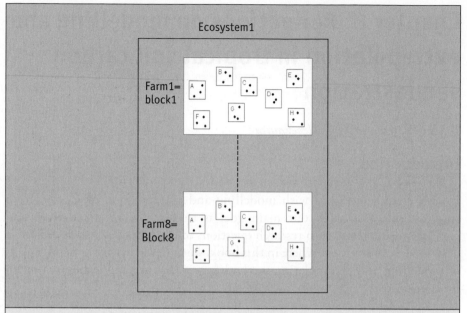

Figure 1. Idealised description of the experimental design, presented for one ecosystem only.

In order to apply a statistical analysis on acquired data, our model assumptions are as follows. Each farmer, a long time ago, partially cleared the Native Forest (NF), and subsequently established a number of plots of more or less equal size and shape to each of which one of the treatments A, B, C, ..., H, was assigned randomly. These plots, together with a similarly chosen plot in the remaining NF part of the farm, form our own long term experiment. The experimental design as described here implies that the cultivated plots have more or less the same age, so that possible differences in soil C stock can be attributed to differences in treatment, and differences in age of establishment are irrelevant.

In statistical terms, the experimental design is a 'complete randomised block design' in which the farms act as 'blocks'. It is 'complete' because all treatments are presened at each farm; 'randomised' because within each block the treatments are randomly assigned to the plots. Under the hypothesis that the farms are randomly selected from the population of farms (with similar properties) in the ecosystem under study, the blocks are so-called 'random blocks'. The treatment NF takes the role of a 'control' or 'baseline'. Due to the experimental design, the plots are the 'experimental units' in a corresponding statistical analysis. In each plot, soil C stock (and other interesting) data were obtained from several horizons of soil pits that were equidistantly situated along linear transects. In order to apply classical statistical analysis following this 'complete randomised block design', the individual C stock measurements should be integrated to one C stock 'observation' per plot, for which a Normal distribution can be assumed.

An additional model assumption is additivity, which implies that the (numerical) difference between the expected values of C stock observation in LMS_i and NF is the same for all blocks. Figures 2A and 2B both elucidate the notion of additivity and present in fact the same information.

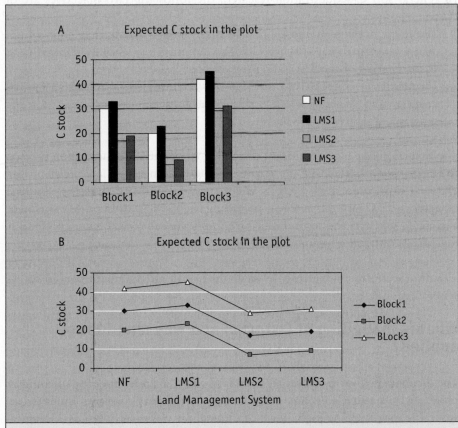

Figure 2. The additivity assumption, in the case of 3 farms (blocks) and 4 LMS (including NF), elucidated in two figures A and B that represent the same information.

One of the major research questions in the C-sequestration project is the identification, per ecosystem, of the 'best LMS', i.e. the LMS that exhibits the highest expected C stock accumulation (which, in Figures 2A and 2B, is LMS_1). Serious violation of the additivity assumption would imply that the 'best LMS' would be farm-dependent, a complexity that impedes the search of the 'best LMS' at the ecosystem level.

The strategy to carry out the identification question *per ecosystem* is in agreement with our earlier observation that one and the same Land Management System (LMS) can exhibit different performance in different ecosystems.

In general soil C-sequestration research, the following two objectives are of central relevance:
1. The identification of those Land Management Systems that have a high (or even optimal) capacity for C sequestration. This is briefly referred to as 'identification of the best Land Management System'.
2. The prediction (forecast) of C stock and/or C accumulation at places unvisited in space (and/or time), which is extrapolation of the results obtained at the experimental farms.

The first objective is an important research goal of the C-sequestration project, to be answered per ecosystem. Although there is an undeniable relationship between both objectives, there is also a fundamental difference, as the first objective needs to be answered by providing the *name(s)* of Land Management System(s), whereas the second objective implies a *numerical value* (preferably including confidence bounds).

The next section will explain why modelling is an indispensable tool in soil C sequestration research.

What is modelling of C stocks, and why should it be applied?

The ultimate goal of modelling is to obtain a better understanding of complex processes in nature, so that we can extrapolate results to places unvisited in space and/or time. The ingredients of a model are chemical/physical/biological knowledge about the process on one hand, and/or datasets of measurements on the process on the other. Datasets obtained from concomitant processes may serve as a useful ingredient.

There are two types of model, which are different in definition and approach. The first mentioned type of model is the so-called *'process based model'*, and in that way is grounded on 'process knowledge'. Models based on the statistical analysis of data sets instead, are called *'probabilistical-statistical models'*. In order to discriminate between systematic and random components in the information contained in the data, such models contain a data reduction device. An important class of models of this type is the well-known statistical regression model. Once the regression coefficients have been estimated (using data reduction), the regression model is ready for forecasting and prediction purposes.

The differences between the abovementioned models are so fundamental that it is difficult or perhaps even impossible to catch the two types of model properly within one common description. For that reason we will use the following definitions.

A *'process-based model'* is a reflection of reality into a mathematical formulation, often translated into a computer program. As a special type, we mention the so-called 'Process Based Simulation model', that step-by-step predicts future process values. As an example of this type of model, we mention the CENTURY model, described by Parton *et al.* (1987, 1993). This type of model is to be discussed in more detail in the next section.

A *'probabilistical-statistical model'* is a data reduction device that extracts information that is viewed as 'relevant'. As an example, we mention the classical analysis of variance ANOVA, in which data reduction takes place via sums and sums of squares, and eventually by one single value: the level of significance of the test, also called the 'p-value'. In relation to the wish to extract information that is viewed as 'relevant', we advocate the use of so-called 'sufficient statistics' which exhaust all the information about the unknown parameter that is contained in the sample. We refer to Hogg *et al.* (2005) for an extensive description of this important type of data reduction. This type of model is dealt with in the sections on classical and spatial statistical approach and pseudo-chronosequence approach.

Probabilistical-statistical models intend to increase quantitative insight in the process, leading to the identification of knowledge gaps, to answer 'what if' questions, and to test hypotheses about possible trends, differences, etc. (Hanson *et al.* 2001).

Models need data. For probabilistical-statistical models this is unmistakably true. In this case, we distinguish two type of data, namely data obtained from the main process, and data from concomitant variables, the latter of which is described by 'explanatory variables', 'regressors', 'factors' and/or 'driving variables'. Statistical regression models are examples of this.

Also process-based models need data, for instance to quantify the transition speed of one phase of the system into another, or, more general, to calibrate (or tune) the process, via parameters (coefficients) in the model. These parameters constitute a semi-permanent part of the model, and in principle should only be adapted if the model is transferred to another 'environment' in which new conditions apply. An example of this is a C sequestration rate that possibly needs to be adapted ('retuned') if the model is transferred from a temperate zone to the tropics. Needless to say that a model should be well-tuned to the environment to which it is applied, and that such tuning should be based on data that reflect 'real life' in this environment.

Apart from semi-permanent data, process-based models usually also need so-called 'input data'. This 'input data', normally offered to the model in a scenario analysis run, quantitatively describe auxiliary processes that influence the model outcome of the process under study. An example is daily precipitation data that (together with other environmental data) is offered to a 'process based simulation model'. We point out that 'input data' usually are an autonomous dataset, not reflecting present reality. Even very unlikely datasets (e.g. excessive precipitation during a long period) could be offered to the model, for instance in order to detect so far unforeseen major implications that a climate change may impose on the process under study (e.g. the C sequestration process). For testing purposes, particularly in the case of long-term experiments, historical data are offered to the model, after which the predictions are compared with field measurements.

A model should be as simple as possible. Of course, there is an area of tension between the simplicity of a model on one hand, and its ability to describe the process on the other. The best model is a model that gives acceptable output in the simplest way (somewhat comparable to 'maximum performance at lowest costs'). The principle that a model description should be as simple as possible ('All things being equal, the simplest solution tends to be the best one') is known under the name Ockham's razor, also called 'law of parsimony', see Ariew (1976).

A point of concern is that, in principle, a model should be used only for what it is meant for. We warn against the widespread practice to apply a model without having any knowledge about its construction and its 'inner side'. In this connection, we strongly advise against 'black-box' use of a model, where one sometimes get the impression that the less the model is understood, the more it is trusted.

The question about the *necessity of a model* can be answered quite briefly. Observations on a process are normally very limited in space and/or time. For prediction, forecasting, and inference objectives at a larger scale, one simply needs modelling. Another warning however is in order here. With respect to a probabilistical-statistical model, experimental results should not be extrapolated too far beyond the experimental conditions. For instance, when a regression is applied with yearly precipitation in the range of 800-1000 mm as an explanatory (measured) variable, one should mistrust model predictions based on a regression equation in which for the precipitation variable is substituted a value of 2000 mm.

For a process-based model we recommend that (at least) it should be fine-tuned when transferred to another environment, i.e. its parameters should be modified according to the new conditions. Unfortunately, it is virtually impossible to modify a (complex) model when one is not the author of it. In computer models it is hardly possible to detect parameters as such, let alone that one has any idea about which numerical value in the computer program should be replaced by which other one.

The choice of a model can have far reaching consequences, in particular in relation to our current topic of C sequestration. This is stated very clearly by Powlson (1996): 'SOM models are also being used to compare the impacts of different land management practices in tropical areas where there is an urgent need to increase primary production in sustainable ways. [...] To use a model in this context that has not been well tested is certainly unscientific. In view of the possible impacts on the lives of people who have no control over the formulation or use of the model it could be also argued that it is immoral'.

Process based simulation (PBS) models

In this section we study the usefulness of Process Based Simulation (PBS) models in answering some important questions in soil C sequestration research. We first start with a description of this particular type of model.

Description of PBS models

Process based simulation models are especially meant to predict soil C stocks and soil C changes, as a function of Land Management System (LMS), climate, topography, soil type and other environmental conditions. Predictions can be obtained not only for (point) locations unvisited in space and/or time, but also for whole regions. Most models pretend to be able to predict future changes. Normally, a vast amount of knowledge has been used to build the model. Nevertheless, there is some doubt about the usefulness and applicability of these models in concrete situations.

Important examples of PBS models are CENTURY, RothC, CANDY, DAISY, DNDC and NCSOIL, see Powlson *et al.* (1996). Most of them predict variables like: long term changes in Soil Organic Matter (SOM) content, soil moisture, soil temperature, plant biomass production, crop yield, nitrate leaching.

Based on input variables (as indicated in the previous section), the processes in the soil-plant system are mimicked, and, depending on the model, the dynamics is simulated in time steps of minutes, hours, up to and including years. For flow diagrams describing the complex dependency in the case of, for instance, CENTURY model, we refer to Parton (1996). In general, calculations concern, among others, C flows from and to various (imaginary) pools, based on various turnover rates, and so on. Parameter estimates and calibration normally depend on local (site specific) data and, quite often, also on 'remote' experience which implies that the model has been calibrated 'elsewhere'. For instance, many PBS models have been calibrated for temperate zones but are applied in tropics.

Unfortunately, no models have been developed with tropical soils in mind due to the fact that research on SOM in the tropics has been much less extensive (e.g. Parton *et al.*, 1989; Shirato *et al.*, 2005) and that long-term data sets are not available for calibration. In the light of increasingly scientific interest in the possible effects of global warming, which seem to manifest themselves more dramatically in vulnerable ecosystems in tropics, we point out that the lack of tailored scientific tools is to be viewed as a severe omission.

The CENTURY simulation model offers a possibility to predict long-term SOC trends based on mathematical representations of C cycling processes in the soil-plant systems. A detailed description of the CENTURY model was published by Parton *et al.* (1987;1993). In practice, a considerable number of different input data is needed in order to run the program. We restrict ourselves to CENTURY (Parton, 1996) and RothC (Coleman and Jenkinson, 1996) which require the following input data, respectively:

For CENTURY, the major input variables for the model include:
- monthly average maximum and minimum air temperature;
- monthly precipitation;
- contents of lignin, N, P, and S in plant material;
- soil texture;
- atmospheric and nutrient inputs;
- amount of plant residues added to the soil;
- vegetation types and CO_2 levels.

For RothC, the input data are:
- monthly rainfall;
- monthly open pan evaporation;
- average monthly air temperature;
- clay content in the top soil;
- an estimate of the decomposability of the incoming plant material, the so-called 'DPM/RPM ratio';
- soil cover – is the soil bare or vegetated in a particular month?
- monthly input of plant residues. As the below-ground part of this input is rarely known, the model is most often run 'in reverse', generating monthly (or more often yearly) inputs from known soil site and weather data;
- monthly input of farm yard manure, if any.

As stated before, input data are used to generate conversion factors, or are converted by parameters inside the model (the semi-permanent data), which together generate predictions.

The usefulness of PBS models

The ability of PBS models to supply predictions for places (regions) unvisited in space and/or time makes is one of their advantages. A main disadvantage is that in general a well-founded quantification of the inaccuracy of the estimates (predictions, forecasts) is lacking. This is fatal for their practical use. If one wants to know the effect of certain policy measures on the C stock, a PBS models is run two times, once with input variables 'as usual', and once with input variables according to the changed 'what if' situation in which the policy measure has been taken into effect. The model generates two – presumably different – values for C stock, but no indication of the error in the estimate is supplied. Therefore, there is no evidence that the policy measure will result in a positive or negative effect compared to the 'status quo' situation. For that reason, PBS models are very limited in appropriate handling of 'what if' questions.

Also in answering the 'identification question', PBS models exhibit a serious deficiency, due to the same lack of quantitative insight in the error. When LMS are to be compared in order to identify the one with the highest C stocks, PBS models could be applied in consecutive scenario analyses where LMSs are successively tried and tested, after which the 'best' would be the LMS that in the simulation run produced the highest C stocks. However, since there is no well-founded insight in any kind of experimental error, we are not able to discriminate between actual 'LMS effect' and 'random fluctuation', so the identification question remains fundamentally unanswered by PBS models. This is one of the main reasons why application of a PBS model in the C-sequestration project has not been considered.

PBS models have some serious additional shortcomings, as follows. Elliot *et al.* (1994) observed that the framework for evaluating regional information is the 'driving variable process property' (DVPP) paradigm that is one of the underpinnings of ecosystem science. The DVPP paradigm implies that any ecosystem is the realisation of a certain fixed number of driving variables that control the whole ecosystem, including flow patterns. It does not account for other variables than the ones included in the list of recognised ones. However, it cannot be excluded that variables that are not included do nevertheless play an important if not crucial role in the process under study.

The consequence of unrecognised and unaccounted for driving variables in the application of PBS models is an unknown and possibly severe deviation in the predictions of the model, so the DVPP paradigm has certainly (but perhaps unavoidably) risky consequences. In environmental research it may be inevitable to apply, in one or another mild form, the DVPP paradigm as a kind of model assumption.

In contrast to its effect on PBS models, non-inclusion of essential factors in statistical hypothesis testing does not affect the validity of the statistical procedure, although it may diminish the power of the test through an (undesirable) increase of the experimental error.

With respect to the DVPP paradigm, Bruce *et al.* (1998) state that 'there appears to be reasonable consensus that state-of-the-art soil organic matter models include the essential factors controlling soil C. [...] Such procedures can build confidence in the internal consistency of the model but do not constitute a strict validation test'. In reply on this, we point out that a 'reasonable consensus' is quite a meagre argument in science where facts instead of consensus should be the decisive factor.

Bruce *et al.* (1998) continue with: 'Many soil C models have been quite successful in simulating C dynamics at the field scale, but they generally require some degree of site-specific calibration. For example, a control treatment in a long-term experiment could be used to calibrate model inputs or initial conditions that are then used to independently simulate other treatments in the field experiment'. Our criticism is two-fold. (1) Soil C models generally pretend to supply predictions (for time horizons of decades to centuries) that cannot be checked with available data sets yet, so the exact meaning of '*quite successful*' remains rather unclear. In this connection, it is tempting to refer to many economic models that impeccably 'predict' a (past) stock exchange crash when applied to historical data, but only quite seldom appear to be able to adequately predict a stock exchange crash that is going to happen in real future. (2) With respect to '*site-specific calibration*', we observe that any '*long-term experiment*' normally goes with some serious difficulties and problems (as we will see in the next paragraph) and are normally even not available, especially in the tropics, let alone that one would have available useful 'site-specific' information derived from a 'long-term experiment' in the direct neighbourhood of the site under study.

With respect to the calibration of PBS models, Powlson (1996) remarks that existing long-term experiments are usually the only source of data. We point out that calibration and testing, using the same long-term dataset, although generally leading to a good match between predicted and truly observed values, would be a proof of irresponsible practice.

With respect to the remark that existing long-term experiments are usually the only source of data, it is regrettable that these 'long-term experiments' and especially the control experiments therein, provide serious problems. We refer to Glendining and Poulton (1996): 'Many long-term experiments were not originally established to measure changes in soil organic matter (SOM) content. Thus, there are inevitable difficulties associated with interpreting SOM data from these sites. The general difficulties associated with long-term experiments mainly arise from the experimental design, sampling methods, and record keeping. These include:
- little or no replication or randomisation;
- lack of time-zero samples;
- inappropriate control treatments;
- incomplete description of the experiment and sampling protocols...'

The authors continue by observing that 'soil samples may not have been taken until the experiment was well under way. Thus,
- the initial SOM content, before the treatments were imposed, may have to be estimated by extrapolation [...].
- lack of time-zero samples may also lead to problems in estimating the rate of change of SOM.'

With respect to application of PBS models in the tropics, we mention the following additional complication, formulated by Van Keulen (2001): 'In principle, organic matter dynamics in tropical (agro-)ecosystems are not different from those in temperate (agro-)ecosystems, but the intensities of the processes are higher, and therefore the rates of turnover are higher. The consequence is that small inaccuracies in the determinants of these rates may have very strong effects on results'.

In practice, PBS models are used as black boxes. This is unavoidable as all their technical ins and outs are far too difficult to be well-understood and well-interpreted by the practical user. Shirato *et al.* (2005 and references therein), remark that for instance CENTURY can produce strongly underestimating or overestimating predictions, in which the prediction may be even several times greater than observed values. In this respect, Parton and Sanford (1989) suggest that some modifications are needed for more accurate simulation in tropical soils, and they describe some additional soil processes that should be taken into account. However, the practice of pointing out factors, the non-inclusion of which should be held responsible for the undesired model outcome, is based on conjectures without any scientific status. The question why these PBS models do not supply a reasonable outcome remains therefore fundamentally unanswered.

We doubt that the solution to overcome these problems should be sought in the direction of an even higher degree of model complexity. Despite their impressive construction, huge pretensions and frills, outcomes of PBS models do not go with any guarantee on accuracy.

Smith *et al.* (1997) evaluated the performance of nine leading SOM models on twelve datasets from seven long-term experiments, all within the temperate region, but inappropriate use of statistical techniques makes their result less trustworthy. Also Smith *et al.* (1996), pretending model evaluation and model comparisons based on quantitative methods, lack any scientific rationale by making a fundamentally inappropriate use of statistical techniques.

Summarising, our objections against the use of PBS models are as follows:
- Error terms are not taken into account, and consequently:
  - predictions are given without any quantitative indication about the corresponding accuracy;
  - trend detection is not possible, nor for C stock changes in time, nor for C stock differences in space.
- Implicit use of initial 'equilibrium assumption' c.q. unclear use of the model when run in 'reverse mode' in order to 'predict' initial conditions as advocated by Coleman and Jenkinson (1996).
- The models are (at least) partially black boxes, and have obscure conversions (e.g. from air temperature to soil temperature).
- The models are normally calibrated and tuned 'elsewhere' and cannot be simply transferred to countries in the tropics.
- The validation and comparison of these models is very poor due to inappropriate use of statistical techniques, implying that a scientifically sound basis for model choice is lacking.

Our final conclusion is that the scientific and practical value of PSB models as applied to soil organic matter is extremely weak. Therefore, conclusions based on model simulations should be used very cautiously and even with the necessary mistrust. In the tropics, where concomitant long-term experiments are frequently lacking, the use PBS models should be avoided, as scientific underpinning is lacking. Science should not take Oracle of Delphi's chair.

The classical statistical approach

In this section, we consider to what extend classical statistical methodology is equipped for dealing with the main objectives of soil C sequestration research as mentioned in the introduction. In order to elucidate the kind of problems that are encountered when classical statistics are applied, we start with the description of a very simple experiment.

Suppose that treatments (e.g. Land Management Systems) A and B are randomly applied to two plots (Figure 3). In each plot, soil C stock measurements are taken by way of soil pits that are randomly located. The goal of the experiment is to obtain statistical information on the difference between A and B with respect to C stock level. The corresponding null hypothesis reads:

H_0: Treatments A and B do not differ with respect to the expected
C stock level.

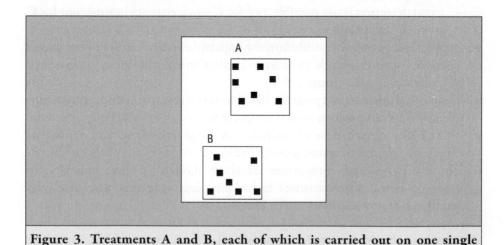

Figure 3. Treatments A and B, each of which is carried out on one single plot. Observations in each plot are obtained from soil pits that are randomly located.

Unfortunately, the experimental design consists of only two experimental units (the plots), far too few to draw any conclusion about any difference in C stock level, so the experimental design is inadequate to detect differences between treatments A and B.

Nevertheless, the application of statistical sampling theory seems to offer a glimmer of hope, a subject treated in much detail by Cochran (1977). Indeed, due to the random allocation of the observations per plot, one is allowed to view the experiment as consisting of two independent samples of independent observations. Under the assumption of Normality, a classical t-test can be applied. As an additional advantage, the power of the statistical test can be increased to any desirable level, by increasing the number of observations (soil pits).

The true situation however, is less favourable than suggested. Plots and treatments are completely confounded, here. That implies that any difference between the two samples might be fully attributable to a difference between the two plots, not necessarily caused by any difference in treatment.

A difference in C stock between two plots is very likely to occur due to natural plot variation. In practical situations, this natural plot variation should be expected to be of considerable importance, which is the reason why application of a t-test as described above leads to rejection of the null hypothesis in case a reasonable number of observations have been obtained from each plot. The subsequent conclusion that treatments A and B differ significantly in C stock, however is completely unfounded. Statistical inference about possible differences between treatments A and B needs to be carried out along other lines, notably another experimental design.

The same arguments hold true in the case of more than two treatments, randomly assigned to the same number of plots, and a random allocation of soil pits in each plot is the experimental design, as illustrated in Figure 4. Under the assumption that the observations have a Normal distribution, a one-way Analysis of Variance (ANOVA) is a permitted statistical tool. In case of a sufficient number of observations per plot, and a relatively high 'between-plots' variation, ANOVA will detect a difference. Again, the conclusion is about (uninteresting) *plot differences*, whereas the results by no means demonstrate any (interesting) *difference between treatments*.

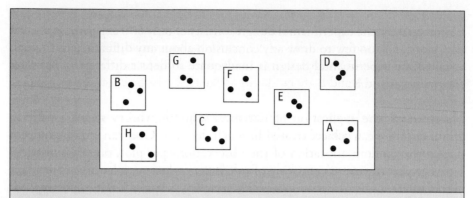

Figure 4. The case of more than two treatments (i.c. treatments A,B, …, H), applied to the same number of plots, where observations within plots are taken from soil pits at randomly selected positions.

With their observation that application of the 'classical F-test' (ANOVA) results in a '(much) too quick' rejection of the null hypothesis, Overmars *et al.* (2003) and Anselin and Griffith (1988) touch upon the problem described here. However, we argue that statistical inference about treatments is *not at all* possible in these cases. It is clear that ANOVA in combination with this type of experimental design is even fundamentally the wrong technique to reveal differences between treatments.

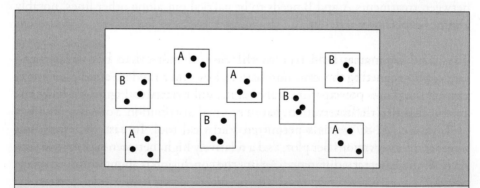

Figure 5. Two treatments A and B, via replications randomly assigned to a number of plots, where observations within plots are taken from soil pits that are located at randomly selected positions.

A solution to the abovementioned problems is the use of replication in the assignment of treatments, such that the 'between-plots' variation can be estimated. A simple example of an appropriate experimental design is shown in Figure 5, where treatments A and B are randomly assigned to a sufficient number of plots.

The statistical analysis in the experimental design shown in Figure 5 can go as follows. The null hypothesis reads:

H_0: Treatments A and B do not differ.

According to the experimental design in which the treatments are randomly assigned to the plots, plots are the experimental units. So, observations within plots should be integrated to plot level, so that each plot is represented by just one real number which is the summary of the observations within the plot. In this new model, the spatial correlation between observations within plots should be taken into account. In the next section, we deal in more detail with spatial correlation and the way to summarise observations within plots.

In the case of two treatments, the statistical analysis follows the classical t-test (applied to the summarised data for which a Normal distribution is assumed) if the variances of these summary statistics are equal, or an appropriate modified t-test in case the variances are unequal. Similar remarks hold for the case of three or more treatments, in which a (modified) F-test can be applied. For a detailed description of suitable statistical techniques in this respect, we refer to Van Putten and Knippers (2007).

Our conclusions concerning the application of classical statistical methods are as follows:
- If replicates of treatments are not available, classical t-test and F-test in sample theoretical setting give the right answer, however to an irrelevant question.
- The statistical power of classical t-tests and F-tests can be high, but thus revealed LMS-effects can be spurious.
- The solution to overcome these problems is an appropriate experimental design including sufficient replications of LMS, preferably in combination with the integration of data to plot level using spatial statistical techniques (to be treated in the next section).
- In case of an appropriate experimental design with a sufficient number of replications, classical statistical methodology can be applied to the summary statistics per plot, resulting in:

- a correct answer to the 'identification question', because quantitative insight in the experimental error can be obtained due to replication;
- the possibility of numerical extrapolation including confidence bounds for predictions.

The spatial statistics approach

In the previous section, we argued that a classical statistical approach for answering important questions concerning soil C sequestration is very useful provided that an appropriate experimental design has been applied in which plot replication plays a crucial role. We pointed out that within each plot the observations need to be integrated to one summary statististic which plays the role of 'one single observation' in a subsequent classical analysis (for instance a two samples t-test or a one-way ANOVA).

In this section, we describe in more detail the process of integration of data per plot to one summary statistic. The formulas presented here are derived from Van Putten and Knippers (2007). In order to describe the procedure, we need to give some details on spatial statistics. For an alternative approach using spatial statistical methodology, we refer to Kravchenko *et al.* (2006). For a general introduction to the subject of spatial statistics, we refer to Chilés and Delfiner (1999) and Cressie (1993) in case a rigorous statistical-oriented approach is desired, and to Isaaks and Srivastava (1989), Davis (2002), Olea (2003) and Fortin and Dale (2005) when practical application is the starting point. For a GIS-related approach and some advanced methodological tools, we refer to Kanevski and Maignan (2004).

An important notion in spatial statistics is 'stationarity', which indicates that expected values of observations and their variances both are constant over the plot and the correlation between two observations is a function of their (vector) distance only. This requirement is roughly fulfilled when the plot is (rather) homogeneous with respect to soil C stock and there is no spatial trend in it. For an impression of the spatial dependency structure of soil C stock in a practical situation, we refer to Heckrath *et al.* (2005).

In a stationary field, the spatial dependency of observations at distance h can be described by means of the so-called (semi-)variogram $\gamma(h)$. It is common practice to describe the dependency by a known parametric (semi-)variogram function. In modelling physical and chemical soil properties the so-called *spherical*

functions seem to be especially appropriate (Odeh *et al.,* 1990; Shukla *et al.,* 2004). Spherical functions have parameters for:
- range (smallest size of the distance *h* such that the semi-variogram remains constant for all distances higher than this value);
- Sill (variance $\sigma2$, the function value in the range);
- Nugget effect (the intercept of the function).

In practice, soil C stock exhibits a high variation and a high range (75-200 m, see McBratney and Pringle, 1998; Kravchenko, 2003; Terra *et al.,* 2004). Figure 6 is a typical picture of a (semi-)variogram function, in which the range is about 80 m, the sill is about 900, and the precise value of the nugget effect has not been indicated (after Kravchenko, 2003).

Given a number *n* of observations, done at locations of which the coordinates are known, the Best Linear Unbiased Estimator (BLUE) of the plot expected value μ is

$$\tilde{Y} = \sum^{n} w_i Y_i$$

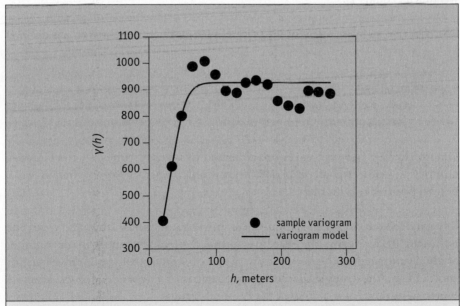

Figure 6. Typical example of a semi-variogram of soil C stock, after Kravchenko (2003).

$i=1$

where the weights w_i can be determined as functions of the various distances of the locations of the observations. The sum of weights equals 1 in order to meet the unbiasedness condition. 'Best' means here that the variance of the estimator takes the minimum value. In a subsequent statistical analysis, classical hypothesis testing is applied to summary statistics.

In a natural landscape, observation points that are close together are more related than widely spaced ones. This implies that close observations are not fully independent and that the amount of information obtained by a specific number of observations may be reduced by their spatial relation. Therefore the effective number of observations (with respect to the information content) may be lower than the actual number. This is worked out in more detail by Van Putten and Knippers (2007).

Let $\hat{\mu} = \sum\limits_{i=1}^{n} w_i Y_i$ be an arbitrary linear unbiased estimator of μ (the sum of

weights w_i equals 1).

The effective number n_{eff} of observations, with respect to estimator $\hat{\mu}$, is defined as

$$n_{eff} = n_{eff}(\hat{\mu}) = \sigma^2 / \mathrm{var}\,\hat{\mu},$$

where σ^2 is the common unknown variance of the observations .

Due to the fact that var $\hat{\mu}$ can be calculated as a factor times σ^2, the effective number n_{eff} does not depend on the (unknown) parameter σ^2, but is just a known positive real number.

Roughly spoken, in terms of equivalent information to be obtained from the estimator under study, the 'effective number' of observations is the number of observations after the dependency of the observations is removed in one or another way. Or, strongly simplified, the number of n dependent observations is 'worth' a number of n_{eff} 'independent' observations.

If the dependency of the observations is positive, which is generally the case in soils, we have $n_{eff} < n$. The higher the dependency rate, the lower is n_{eff} with limiting case $n_{eff} = 1$. For the opposite case in which the observations are independent, we have $n_{eff} = n$, which follows directly from basic statistical calculations.

The 'optimal allocation of observations', given the total number of observations, is a question of very practical importance. In Figure 7, in the case of 4 observations, some examples of allocation are given. Corresponding effective numbers are presented in Table 1 for various values of the range parameter varying from 'complete independency' to 'full dependency'. From Table 1, it can be deduced that in the case of 4 observations, the 'corners' allocation is consistently the

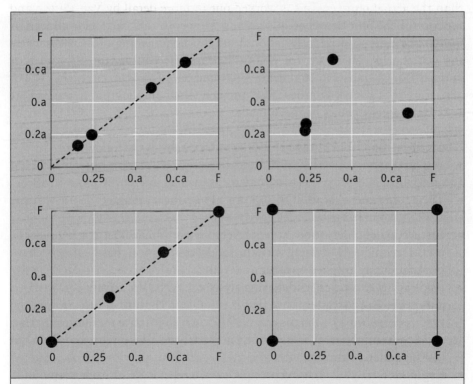

Figure 7. Some examples of allocation of soil pits in a square plot. Clockwise: random along transect, random in plot, in corners, and equidistant along transect.

Table 1. Effective number of observations for a spherical model in which r is the range parameter (the higher r, the higher the dependency).

n_{eff} with $n=4$	Random along transect	Random in plot	Equidistant along transect	Corners
$r=0$	4	4	4	4
$r=1.2$	1.8	1.9	2.4	3.7
$r=1.7$	1.5	1.6	2.0	2.7
$r=2.6$	1.3	1.3	1.6	1.8
$r=5.3$	1.1	1.1	1.2	1.3
$r='\infty'$	1	1	1	1

superior of the four described allocations. However, this particular allocation seems to be quite sensitive to deviations from the model assumptions, particularly the stationarity. Therefore, for practical purposes, 'equidistant allocation of samples along a transect' could be the preferred one. This is precisely the type of allocation that was used in the CSEQ project, which implies that in the project a very appropriate choice has been made.

Concerning the spatial statistical approach we establish the following.
- A spatial approach with a considered allocation of soil pits – preferably even an optimal allocation – is necessary for statistical inferences about LMS.
- The equidistant allocation of soil pits along linear transects as used in the C-seq. project has been a very good choice.
- Integration of observations at plot level to one summary statistic per plot is a useful technique for dealing with the model assumption that *treatments* have been randomly assigned to plots.
- Application of classical statistical analysis, e.g. ANOVA, to plot-summarised data is possible. An alternative that is suitable to the aim of 'selection of the best' is described by Gibbons *et al.* (1977). An alternative methodology that avoids making Normal assumptions is described in Lehmann and Romano (2006) and in Wasserman (2007).
- Regression analysis, e.g. using precipitation, temperature, altitude, slope, and/ or time as explanatory variables, is a possible generalisation of what has been discussed in the current section.
- Prediction and forecasting, including confidence bounds, is possible.

The pseudo-chronosequence approach

A (real) chronosequence, (only) in principle would be useful to measure the effect of a specific factor, such as a land management system, over time. All factors except time are kept constant. If we want to describe the development of C stocks in grassland that is established after clearing the native forest, the ideal way to obtain a chronosequence would be to describe, in the same plot, the stocks in the native forest, immediately after clearance, and further in time increments until a new equilibrium has been established. This approach is unpractical for two main reasons: (1) it is too time-consuming and costly, and (2) equilibrium will not be reached even in a century.

In areas with homogeneous soils, it is possible to use a pseudo-chronosequence instead. With the forest-grassland conversion, such a pseudo-chronosequence can be obtained if the conversion of native forest to grassland in closely related fields occurred at different moments in time, and the grassland in all converted plots was managed the same way until the moment the observations are made (at absolute time T). For each plot, the 'age' is determined, i.e. the time (in years) that has elapsed since the native forest was cleared and was replaced by grassland (Figure 8, T is 2007).

A necessary model assumption is that the C stock under native forest was originally equal in all converted plots and in equilibrium. Additionally we assume that the C stock follows a type of (decay) function from the moment that the plot has become converted onwards, in a functional form that is the same for each plot independent of the moment of conversion. In case of bad management of the grassland, the decline in C stock would be quick and heavy, but in a well-managed improved grassland there may be a slow increase. Both decline and increase would be asymptotical. In the following, we will speak of a 'decay' function, even if the C stock increases.

We will use an exponential decay function: $f(t) = \alpha + \beta\exp(\gamma\,t)$, which is the solution of the simple differential equation for C stocks that was formulated by Hénin and Dupuis (1945), where t is the age of the plot. Parameter γ is negative always, and β is positive in case of a decrease in soil C stocks and negative in case of increase. In the (exceptional) case that the conversion to grassland would not affect the C stocks as they were reached under native forest, parameter β equals zero, whereas parameter γ in that very particular case remains undefined. Parameters α, β and γ depend on the treatment.

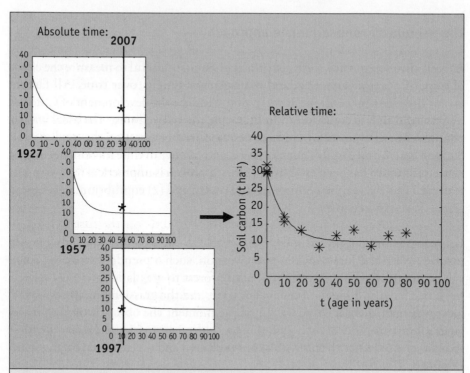

Figure 8. Schematic impression of a (pseudo-)chronosequence in the case of (only) one treatment. The left hand side pictures show observations from different plots (only 3 are depicted) and their (constant) model decay function. The observations and their common decay function are brought together in the right hand side picture (and more observations are added to give a clearer idea about the connection between model function and observations done at various stages in the process).

The second model assumption is that the observations (all made at time T, but representing different ages of the grassland), have as an expected value $f(t) = \alpha + \beta \exp(\gamma t)$ where t is the age of the plot. Thirdly, a Normal distribution of C stocks with equal variances for the (not yet integrated) observations per plot, together with a constant between-plots variance, appears to be a reasonable additional model assumption.

In the model, parameter α takes the role of (expected) asymptotic value of soil C stock in the grassland. The (expected) value of C stock for native forest in equilibrium phase is equal to $\alpha + \beta$, so that β is the expected loss in C stock after

Carbon sequestration in tropical grassland ecosytems

establishment of the grassland. Parameter γ rules the decay rate. A highly negative value of γ would imply that the change to the new equilibrium phase induced by grassland evolves very quickly (Figure 8). We refer to for a schematic impression of the basic idea of the pseudo-chronosequence, where T is the year 2007.

Following standard statistical techniques, parameters α, β and γ can be estimated from the data. These estimates go with a quantitative measure of accuracy, so that confidence intervals for the parameters can be given. This is an advantage of the statistical model as described here, when compared to the common practice of using PBS models in pseudo-chronosequences.

The quality of the corresponding estimators depends on the distribution of the observations, i.e. on the relative moment the observations have been taken. For instance, if all observations come from plots with a quite young age, only the parameter $\alpha + \beta$ (initial value of the NF) can be estimated with reasonable accuracy. If, in the opposite case, all observations come from plots with a quite 'old age', only the parameter α (equilibrium value after treatment A) can be estimated with reasonable accuracy. In practice, a good (if possible: optimal) experimental design therefore is very important.

In practice, different *consecutive* 'treatments' may have been used on the same field. This seems to invalidate the 'pseudo-chronosequence' concept. In the following we will briefly show how such a situation can be used in a model we call a 'consecutive pseudo-chronosequence'. Suppose the plots involved in our study have been treated consecutively with one or more treatments out of a fixed list of treatments A, B, C, ..., not necessarily in that order. Furthermore, suppose that for each treatment is known during which (absolute) period the treatment has been applied. At a certain (absolute) time moment T, the 'present', C stock observations have been obtained for all plots involved and have been integrated to one summary statistic via techniques described in the previous section.

For an impression of how the evolution of the soil C stock in one certain plot could take place, we refer to Figure 9.

When modelling this situation, we describe the evolution of C stock as a function of time using consecutive exponential decay functions with parameter sets α, β and γ that depend on the treatments involved and on the C stock level at the start of the treatment. As a 'shock' in C stock level cannot be excluded just in the very short time that the one treatment is replaced by the next, also a 'shock parameter'

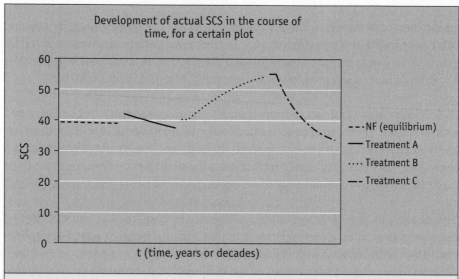

Figure 9. A possible development of soil C stock (SCS) in a plot in the course of time, in response to different treatments (in a hypothetical permanent soil pit).

can be added to the model, describing the sudden change in expected C stock level when one specific treatment is replaced by another (the shock parameter may depend on the treatments involved). We refer to Watson *et al.* (2000), where pictures on p. 74, 209 and 210 give an impression about how C stock can evolve in time under different consecutive land management systems.

We assume error distributions in observations similar to those given in the case of one single treatment A only. If sufficient plots are involved and sufficient diversification is present in the application (and order) of the various treatments, all parameters $\alpha_A, \beta_A, \gamma_A, \alpha_B, \beta_B, \gamma_B, \ldots$ and their confidence intervals, including the 'shocks', can be estimated. It is possible to add explanatory variables such as slope and precipitation, for the effect of which regression techniques can be used.

It is not strictly necessary to do all observations at exactly the same absolute time T. In practice, it is sufficient to make the observations within a reasonably short time interval (of, say, weeks or perhaps even months) as the pseudo-chronosequence analysis will not really be affected by a minor deviation of this kind.

If sufficient historical information is available, the 'consecutive pseudo-chronosequence' technique as proposed here looks rather promising. Implications for various experimental designs have been worked out by Van Putten and Van Wijk (unpublished data), who also give a sensitivity analysis.

A 'consecutive pseudo-chronosequence' has the advantage that it does not require many years, or even decades, of data collection, which from a practical point of view is not feasible. Moreover, a 'consecutive pseudo-chronosequence' is appropriate to estimate just those parameters that really matter in practice.

Another aspect is that no long term contracts with farmers are needed, but only their consent for onetime observations in their land management systems, and their help in describing the history of the plot(s): which treatment was carried out during which period after the native forest was cleared. It is a challenging idea that, in this way, knowledge of the history implies knowledge of the soil C sequestration process in all desired aspects. The problem of uncertainties in historical plot information can be overcome by means of appropriate mathematical technology.

A point that needs some special attention is the following. In the introduction of this Chapter, we remarked that an important research question is the identification of the LMS that has the highest capacity for C sequestration. In the framework of the C-seq. project, where the plots under study were supposed to be (about) the same 'age', it is simply that the LMS that gives the highest C accumulation (compared to natrive forest) during the (fixed) time period that equals the 'age' of the plots. However, in the more general situation of a variable time of treatment, there is some ambiguity in the definition of what is 'the best' LMS. A LMS that has a very high C stock equilibrium level but takes a long time to reach this equilibrium, could rightly be considered as 'less good' when compared to a LMS that reaches a somewhat lower C stock level but in a relatively short time. By a careful description of the parameters it is possible to obtain an appropriate ordering in LMSs which defines when the one LMS is considered to be 'better' than the other. The other important goal in C sequestration research, which is (numerical) extrapolation of experimental results, is possible in principle as well, including confidence bounds for predictions and forecasts.

Extrapolation in conclusions and recommendations

In general soil C sequestration research, two objectives are of central relevance, namely the identification of the best LMS, and the prediction (forecast) of C stock at places unvisited in space (or time).

These objectives are clearly different, as identification of the best LMS requires a name, whereas prediction (forecasting) asks for a numerical outcome. Both, however, require extrapolation of experimental data. Consequently, modelling in one or another form is an essential tool for both objectives.

PBS models are able to deal with extrapolation when numerical prediction and/ or forecast are to be produced. When the unvisited location (or area) has been described through relevant input data (of spatial and/or temporal type), a run of the PBS model gives an outcome. However, PBS models do not generally give any insight in the accuracy of the outcome, so in the case of numerical extrapolation they are of very limited use.

In order to answer the 'identification question', a PBS model could be run a number of times for different LMSs, where input data (as far as not depending on the LMS under study) are kept the same. The PBS model produces soil C stock predictions for each of the LMS under study. Thus, the best strategy for answering the 'identification question' would be to give the name of the LMS that, in the simulation run, produces the highest value of soil C stock over a given time. However, as no insight is given in the accuracy of the predictions, the identification question remains fundamentally unanswered. Any statement about what is 'the best' LMS is given without a quantification of the confidence level.

An additional problem, already referred to above, is that PBS models are specifically meant for use in unvisited locations, while there is no guarantee that the process under study exhibits behaviour that is analogous to the location where the PBS model was tested and tuned. PBS models assume the driving variable process property (DVPP) paradigm, which is a rather heavy model assumption.

So-called 'probabilistical-statistical models', of which the statistical regression models are the most prominent, can in principle answer both type of extrapolation questions, as follows.

With respect to the 'identification of the best', classical statistical analysis applied to plot-integrated data supplies an appropriate framework to answer the question. Statistical testing of hypothesis, preferably carried out via so-called 'multiple comparison procedures' can identify 'the best LMS', and provide quantitative insight in the experimental error. We refer to Hochberg and Tamhane (1987) for details. Although the technique to select 'the best' treatment via these 'multiple comparison procedures' is quite commonly used, it can be viewed as a labourious way compared to the statistical methodology of 'ranking and selection of populations' that supplies the answer in a more direct and elegant way (see Gibbons *et al*, 1977, for more details). It can be shown that selection of 'the best' treatment via the ranking and selection of population techniques can be written as a special form of the multiple comparison procedure, so in our case both methodologies lead to the same result.

If numerical extrapolation is used, the regression model is ready for prediction purposes once the regression coefficients have been estimated (using appropriate data reduction), and a quantitative measure of the accuracy is available. In case the factor time is included as an explanatory variable in the regression model, also a forecast of soil C stock is possible. However, one should be extremely cautious in prediction of values that are far beyond the experimental region, i.e. using values of explanatory variables that are remote from the experimental range. Forecasting with a far horizon is always a risky enterprise. Also, a combination of two explanatory variables can be far remote from the (2-dimensional) experimental region of that combination (e.g. if explanatory variables 'temperature' and 'incoming radiation' tend to occur in combination 'high-high' and 'low-low', a combination 'low-high' is far remote from the 2-dimensional experimental region of the combination of the two explanatory variables). In the latter case, predictions based on an 'unusual' combination of values of explanatory variables should be treated with the necessary caution, as well. This topic is related with robustness of model assumptions and with sensitivity analysis. In this connection, we refer to Saltelli *et al.* (2004).

A remarkeble difference between PBS models on the one hand and statistical models on the other hand, is the following. Non-inclusion of potentially essential factors, in PBS models can lead to (severe) deviations in the predictions, thus leading to wrong results (see also Parton *et al.*, 1989). In contrast, non-inclusion of potentially promising explanatory variables generally results in less accurate predictions due to higher variance (so, leading to less narrow confidence intervals), but the validity of the statistical procedure remains unaffected.

In the C-sequestration project, the application of statistical methodology following procedures as described above has been appropriate for the 'identification of the best' as well as in dealing with numerical extrapolation. In the C-sequestration project, the 'identification question' was considered the most important one. Identification of the 'best' was carried out per ecosystem, for reasons given in the introduction of this Chapter. While the C-sequestration project was carried out in Colombia and Costa Rica 'only', the question of up-scaling the results to similar regions in Tropical America was important. Countries with similar ecosystems are Ecuador, Peru, Bolivia, Panama, Nicaragua, Honduras, El Salvador, Guatemala, and Brazil, but the main question is which limits should be used in extrapolation. Up-scaling is a common research question in applied research, as 'local' field experiments are normally carried out with the ultimate goal to use the results a region as large as possible. Up-scaling is also a kind of extrapolation, but it needs a very careful treatment because the up-scaling concerns regions that were not considered at all in the original research. They are geographically far removed from the locations were the observations were done, and the large geographical distance could, more stealthily, imply environmental changes influencing soil C sequestration processes. If we consider the extrapolation within an (*ex post* constructed) statistical regression framework, we have to define explanatory variables that really matter in this particular case. Within the C-seq. project, it was decided that 'Agri-Ecological Zone (AEZ)', 'altitude', 'slope' and 'soil type' would serve as a reasonable choice of explanatory variables. Although distance to the experimental sites is not accounted for in this particular regression, it is clear that extrapolations to remote unvisited locations are risky. For that reason, it was decided that only 'naïve predictions' would be carried out, i.e. experimental results were declared to be valid for another location in Tropical Latin America only if the aforementioned four explanatory variables for that particular location, coincided (at least to a high degree) with the values at the research site with which it was compared.

Although, given the choice of the explanatory variables, this is the safest way of extrapolation, the underlying paradigm of driving variable process property (DVPP) remains the fundamental basis of the extrapolation, with the aforementioned 'Agri-Ecological Zone (AEZ)', 'altitude', 'slope' and 'soil conditions' are the 'driving variables' in this particular case. In this connection, up-scaling of conclusions needs to be handled very prudently. This is equally true for the extrapolation (up-scaling) of the question of which is 'the best' LMS.

Results obtained from 'consecutive pseudo-chronosequence' research can in principle be up-scaled via similar techniques, with the same caveat.

Chapter 9. Extrapolation of results to similar environments in Tropical America

V.W.P. van Engelen and J.R.M. Huting

Introduction

The four ecosystems of the C-seq. research sites – (1) the tropical Andean hillsides in Colombia, (2) the humid tropical forest, Amazonia in Colombia, (3) the humid tropical forest at the Atlantic coast of Costa Rica, and (4) the sub-humid tropical forest at the Pacific coast of Costa Rica – are representative of larger areas of the continent. This chapter describes an extrapolation study in tropical America of similar environmental conditions as found in the four ecosystems. Such areas were identified in:

1. Colombia and the neighbouring countries of Ecuador and Peru for ecosystem 1.
2. Colombia and the neighbouring Amazonian countries of Brazil, Ecuador, Guyana, Peru and Venezuela for ecosystem 2.
3. Costa Rica and the neighbouring countries of Panama, Nicaragua and Guatemala for ecosystems 3 and 4.

Similarity was based on climatic parameters, topography (elevation and slope) and soil conditions. The maps were derived from an analysis of the Global Agro-Ecological Zones Map – AEZ (FAO-IIASA, 2000), the SRTM90 digital elevation model (USGS, 2003) and the Soil and Terrain Database for Latin America and the Caribbean – SOTERLAC (Dijkshoorn *et al.,* 2005; FAO *et al.,* 1998) for the region. The extrapolation for the Amazon ecosystem was based on the analysis of AEZ and SOTERLAC only, but a subdivision on slopes was made to distinguish the undulating zones from the flat areas. The available (sub)continental data allowed only for a low resolution of the maps produced (scale 1:5 million).

Characterisation of the ecosystems

Agro-ecological conditions of the four ecosystems were defined in broad terms by this project (C-sequestration) (Chapter 1). To allow a geographical extrapolation to similar environments the four ecosystems were characterised in terms of their broad agro-ecological conditions using the Global Agro-Ecological Zones project – GAEZ (FAO-IIASA, 2000). The GAEZ map shows 15 Length of Growing Period (LGP) classes of 30 days intervals (see Figure 1).

LGP has been calculated by GAEZ with the following assumptions and definitions:
1. water balance for a standard FAO reference crop;
2. soil with 100 mm of Available Water Content (AWC);
3. average climatology data from the Climatic Research Unit of the University of East Anglia for the years 1961-1990 (0.5 degrees) interpolated to 5 arcminutes.

The ecosystems of the research sites were further characterised by elevation and slope, both available from GPS data collected by C-sequestration at the sites of ecosystems 1 and 2. Missing elevation data and slope gradient of ecosystems 3 and 4 were derived from the global SRTM90 Digital Elevation Model – DEM (USGS, 2003) (Figure 2).

Soil characterisations of the research sites were based on the work of C-seq. Classification was according to the US Soil Taxonomy (Soil Survey Staff, 1999). Selected analytical data were available for some of the profiles (soils of the Andean Hillsides and the Amazonia ecosystems) focusing on organic matter related data. Morphological description data, however, were lacking. Descriptions and analytical data from the Costa Rica sites were not available. As the spatial soil information – SOTERLAC – follows the Revised Legend of the Soil Map of the World (FAO *et al.*, 1988), the soil names had to be reclassified to the Revised Legend. The lack of standard profile descriptions and the limited analytical data only permitted a rough reclassification.

Spatial soil data were derived from the Soil and Terrain Database of Latin America and the Caribbean at scale 1:5 million (Dijkshoorn *et al.*, 2005). Mapping units contain information on the terrain units and their soil components. Each mapping unit (SOTER unit) consists of up to three soil components of which the proportional percentage (relative area) is given. Ranges of ecosystem attributes are shown in Table 1.

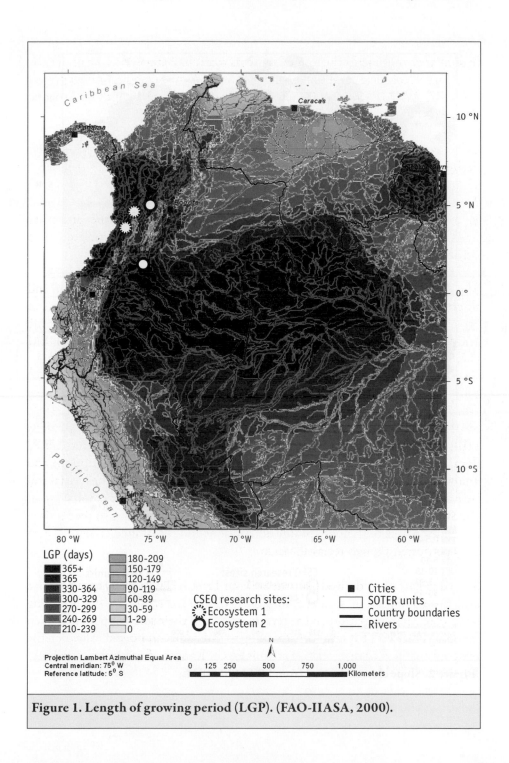

Figure 1. Length of growing period (LGP). (FAO-IIASA, 2000).

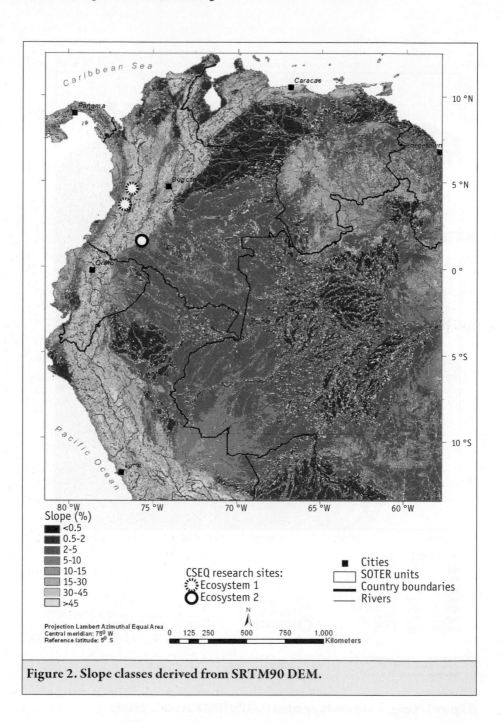

Figure 2. Slope classes derived from SRTM90 DEM.

Table 1. Selected ecological conditions of the four Carbon sequestration ecosystems (Chapter 1; FAO-IIASA, 2000; USGS, 2003).

Ecosystem	Site	Soils	LGP	Altitude (m.a.s.l.)	Slope (%)
1	Dagua	Typic Dystropepts	2	1300-1400	15-45
1	El Dovio	Typic Dystrandepts	2	1700-2000	35-65
2	La Guajira	Typic Kandiudults, Paleudults	1	200-300	2-5
2	Pequín	Typic Paleudults	1	200-300	10-15
2	Santo Domingo	Oxic Dystropepts	1	200-300	2-5
2	Balcanes	Typic Paleudults, Hapludults	1	200-300	10-15
3	Pocora	Inceptisols	1	200-300	0-5
4	Esparza	Inceptisols/ Entisols	4	200-300	15-30

Extrapolation method

Selection criteria of areas with similar characteristics for each of the four ecosystems were:

1. A selection of the LGP class corresponding to the C-seq. ecosystem site (Figure 1).
2. An overlay of the map resulting from (1) with the elevation range representing the ecosystem: this map was derived from the SRTM90 DEM.
3. An overlay of map (2) with the slope class map derived from the SRTM90 DEM using the slope ranges observed within the ecosystem (Figure 2).
4. An overlay of map (3) and the SOTER unit map with the condition: soil classification of the soil component of the SOTER unit = soil classification in the selected research site.

The maps used in steps 1, 2 and 3 were in raster format. Cells' sizes vary between 90 m for the SRTM90 DEM to 5 arcminutes (about 9×9 km at the equator) for the LGP map. The LGP map has been resampled to the pixel size of the SRTM90 to make the resulting overlay less blocky, while maintaining the resolution of the LGP map. Selection of the appropriate LGP, slope and elevation values was done by a GIS query for each ecosystem. The resulting raster map was vectorised and subsequently combined with the map showing SOTER units that have soils similar to those of the considered ecosystem.

Extrapolation criteria

The extrapolation criteria for each ecosystem are described below, and summarised in Table 2.

Table 2. Criteria for extrapolation of research sites characteristics.						
Ecosystem	Site	Soil name[a]	%[b]	LGP[c]	Altitude (m)	Slope (%)
1	Dagua	Umbric Andosols	75	1	1,200-2,000	>10
1	El Dovio	Dystric Cambisols	25	1	1,200-2,000	>10
2	both	Haplic Acrisols/	70	1' and 1	< 300	0-5
		Ferralsols	30			
3	Pocora	Dystric Cambisols	50	1	< 300	0-5
4	Esparza	Cambisols: Umbric	50	4	< 300	15-30
		Dystric	25			
		Eutric	25			

[a]Soil classification from the SOTERLAC database (Dijkshoorn *et al.*, 2005).
[b]Relative area of the soil with the SOTER unit.
[c]LGP 1'= 365+ days, LGP 1 = 365 days, LGP 4 = 270-299 days.

Ecosystem 1, Andean Hillsides, Colombia

The two sites – Dagua and El Dovio – both fall within the LGP class 1: 365 growing days. GPS-measured altitudes range between 1,439 and 1,899 m for the Dagua site and between 1791 and 1859 m for the El Dovio site. A range of 1,400-2,000 m was used for the extrapolation. Slopes at the sites were described as 'moderate and steep' for Dagua and El Dovio respectively (Chapter 1), while the DEM quantifies them as 6-16% for Dagua, with a 30-45% outlier in one site, and 3-30% in El Dovio. Slopes greater than 10% were used for the extrapolation.

Soils of El Dovio were classified as *Typic Dystropepts* corresponding to *Dystric Cambisols* of the Revised Legend (FAO *et al.*, 1988). The soils of the Dagua sites were *Typic Dystrandepts* and correlated with *Umbric Andosols*. According to SOTERLAC, the research sites fall in a SOTER unit with two soil components: 70% *Umbric Andosols* and 30% *Dystric Cambisols*.

Ecosystem 2, Humid Tropical Forest, Amazonia, Colombia

All sites in this ecosystem fall within LGP zone 1: 365 growing days with year-round excess moisture conditions. Also LGP zone 1: 365 growing days was used for the extrapolation. Altitudes for the four sites in the ecosystem range from 232 to 244 m according to the DEM. An altitude of less than 300 m was used in the extrapolation. Slopes of the sites range from 2-15% as determined by the DEM, while the site descriptions give two ranges of slopes from 0 to 5% and from 10 to 15%. The latter two ranges were used in the extrapolation.

Soils range from *Typic Kandiudults, Paleudults* and *Hapludults* to *Oxic Dystropepts*. Tentatively, they were correlated with *Haplic Acrisols* and *Ferralsols*. According to SOTERLAC, the research site occurs in a unit with 70% *Haplic Acrisols* and 30% *Haplic Ferralsols*.

Ecosystem 3, Humid Tropical Forest, Atlantic Coast, Costa Rica

All sites in this ecosystem fall within LGP zone 1: 365 growing days. Zone 1 was used for the extrapolation. Altitudes for the sites in the ecosystem range from 25 to 190 m in the DEM. An altitude of less than 200 m was used in the extrapolation. Slopes of the sites range from 0-5% as determined by the DEM. These percentages have been used for the extrapolation.

Soils of the sites were classified by C-seq. as *Inceptisols*. Tentatively, they were correlated with *Dystric Cambisols*. The spatial information of SOTERLAC indicates for the site a unit with 50% *Dystric Cambisols*, 25% *Eutric Gleysols* and 25% *Dystric Fluvisols*. of which only the first soil type was used in the extrapolation.

Ecosystem 4, Sub-humid Tropical Forest, Pacific Coast, Costa Rica

All sites in this ecosystem fall within LGP zone 4: 270-299 growing days. This zone was used for the extrapolation. Altitudes for the sites in the ecosystem range from 25 to 300 m in the DEM. An altitude of less than 300 m was used in the extrapolation. Slopes of the sites range from 15-30% as determined by the DEM. These percentages were used for the extrapolation.

Soils of the sites were classified as *Inceptisols* and *Entisols* by C-seq. corresponding with Cambisols of the Revised Legend. The SOTER unit that covers the research area comprises *Cambisols*: *Umbric* (50%), *Dystric* (25%) and *Eutric* (25%).

Results

Ecosystem 1, Andean Hillsides, Colombia

An area of about 7,000 km^2 has *Umbric Andosols* and 3,000 km^2 *Dystric Cambisols* with identical LGP and slope/elevation conditions as those observed in ecosystem 1 (see Table 3, Figures 3 and 4).

Table 3. Areas (x1000 km2) for the soil types of ecosystem 1, by country.

Country	Umbric Andosols	Dystric Cambisols
Colombia	6	2
Ecuador	<1	<1
Peru	-	<1
Total	7	3

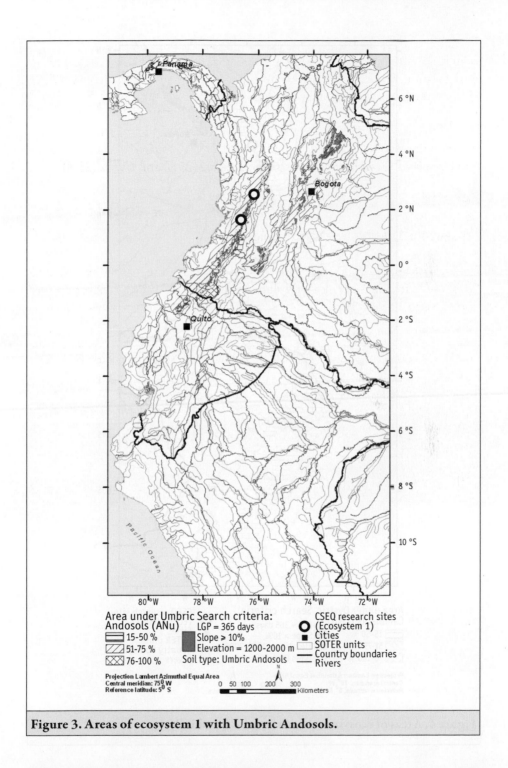

Figure 3. Areas of ecosystem 1 with Umbric Andosols.

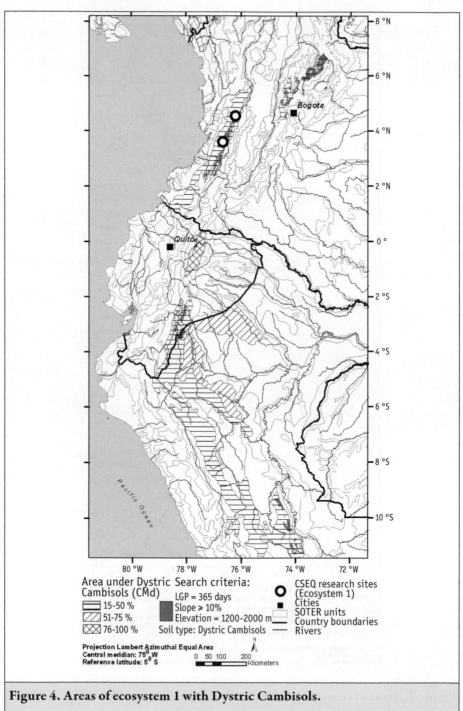

Area under Dystric Search criteria:
Cambisols (CMd)
☐ 15-50 %   LGP = 365 days
▨ 51-75 %   ■ Slope ≥ 10%
▨ 76-100 %   Elevation = 1200-2000 m
              Soil type: Dystric Cambisols

○ CSEQ research sites
  (Ecosystem 1)
■ Cities
☐ SOTER units
── Country boundaries
── Rivers

Projection Lambert Azimuthal Equal Area
Central meridian: 75°W
Reference latitude: 5°S

N
0 50 100   200
              Kilometers

Figure 4. Areas of ecosystem 1 with Dystric Cambisols.

Ecosystem 2, Humid Tropical Forest, Amazonia, Colombia

This ecosystem is extensive in the upper Amazon basin in the undulating to flat lowlands. Soils in these areas are mainly *Haplic Acrisols* and *Haplic Ferralsols*.

Some 288,000 km^2 are *Haplic Acrisols* and 80,000 km^2 are *Haplic Ferralsols* with similar LGP and slope/elevation conditions (flat topography) as ecosystem 2 (Figures 5 and 6, and Table 4). Rolling topography occurs in minor areas in ecosystem 2 (Figures 7 and 8 and Table 5).

Table 4. Areas (x1000 km2) for the soil types of ecosystem 2 on flat topography, by country.

Country	Haplic Acrisols	Haplic Ferralsols
Brazil	111	19
Colombia	112	60
Ecuador	4	-
Guyana	-	1
Peru	60	-
Venezuela	<1	-
Total	288	80

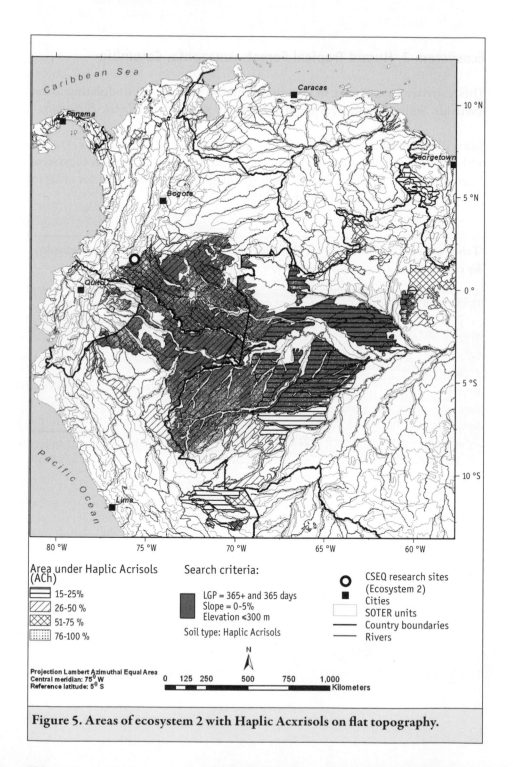

Figure 5. Areas of ecosystem 2 with Haplic Acxrisols on flat topography.

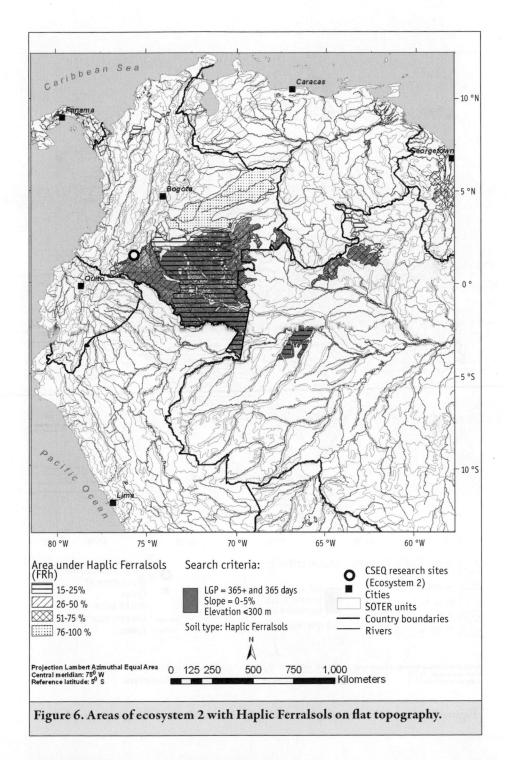

Figure 6. Areas of ecosystem 2 with Haplic Ferralsols on flat topography.

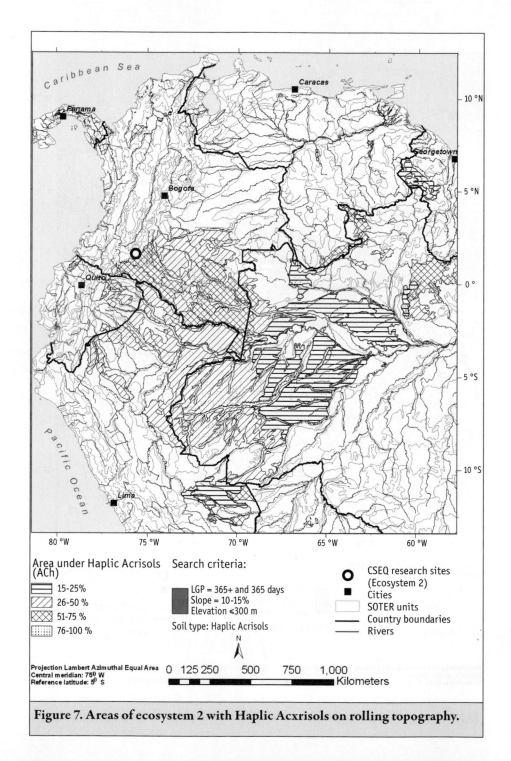

Area under Haplic Acrisols (ACh)

▤	15-25%
▨	26-50 %
▧	51-75 %
▦	76-100 %

Search criteria:

■ LGP = 365+ and 365 days
Slope = 10-15%
Elevation ≤300 m

Soil type: Haplic Acrisols

○ CSEQ research sites (Ecosystem 2)
■ Cities
☐ SOTER units
— Country boundaries
— Rivers

Projection Lambert Azimuthal Equal Area
Central meridian: 75⁰ W
Reference latitude: 5⁰ S

0 125 250   500   750   1,000
Kilometers

Figure 7. Areas of ecosystem 2 with Haplic Acxrisols on rolling topography.

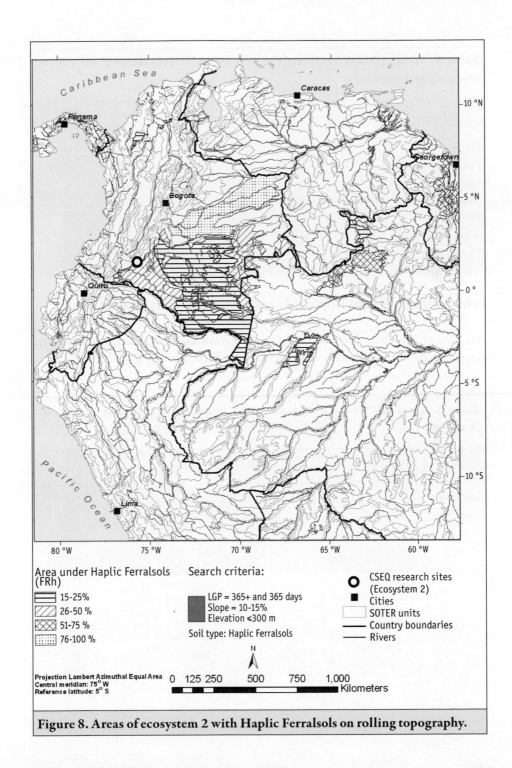

Area under Haplic Ferralsols (FRh)

- ▬ 15-25%
- ▨ 26-50 %
- ▧ 51-75 %
- ▒ 76-100 %

Search criteria:

■ LGP = 365+ and 365 days
Slope = 10-15%
Elevation <300 m

Soil type: Haplic Ferralsols

○ CSEQ research sites (Ecosystem 2)
■ Cities
☐ SOTER units
— Country boundaries
— Rivers

Projection Lambert Azimuthal Equal Area
Central meridian: 75° W
Reference latitude: 5° S

0 125 250   500   750   1,000
Kilometers

Figure 8. Areas of ecosystem 2 with Haplic Ferralsols on rolling topography.

Table 5. Areas (x1000 km2) for the soil types of ecosystem 2 on rolling topography, by country.

Country	Haplic Acrisols	Haplic Ferralsols
Brazil	3	<1
Colombia	3	2
Ecuador	<1	-
Guyana	<1	<1
Peru	1	-
Venezuela	<1	-
Total	7	2

Ecosystem 3, Humid Tropical Forest, Atlantic Coast, Costa Rica

An area of less than 750 km^2 has *Dystric Cambisols* with identical LGP and slope/elevation conditions as ecosystem 3 (see Table 6 and Figure 9).

Table 6. Areas (x1000 km2) under Humid Tropical Forest (ecosystem 3), by country.

Country	Ecosystem 3
Costa Rica	0.7
Guatemala	<0.1
Panama	<0.1
Total	0.7

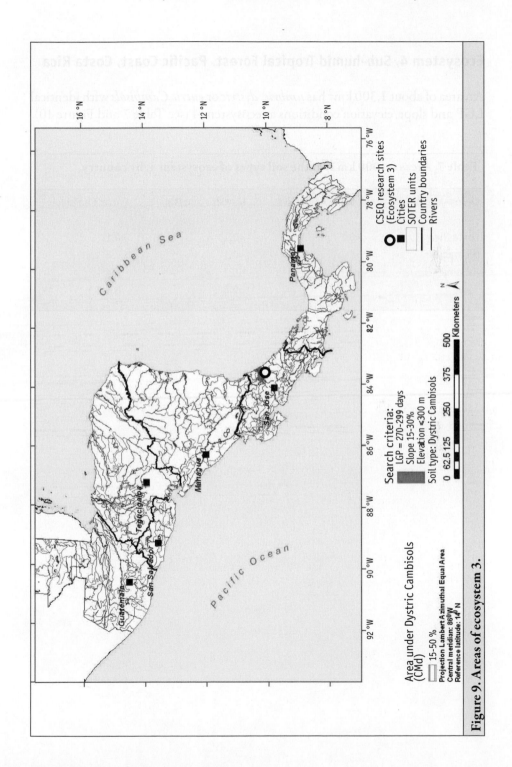

Figure 9. Areas of ecosytem 3.

Ecosystem 4, Sub-humid Tropical Forest, Pacific Coast, Costa Rica

An area of about 1,300 km^2 has *umbric, dystric* or *eutric Cambisols* with identical LGP and slope/elevation conditions as ecosystem 4 (see Table 7 and Figure 10).

Table 7. Areas (x1000 km2) of the soil types of ecosystem 4, by country.			
Country	Umbric Cambisols	Dystric Cambisols	Eutric Cambisols
Costa Rica	<0.1	<0.1	<0.1
Nicaragua	-	<0.1	-
Panama	<0.1	0.7	0.4
Total	<0.1	0.8	0.4

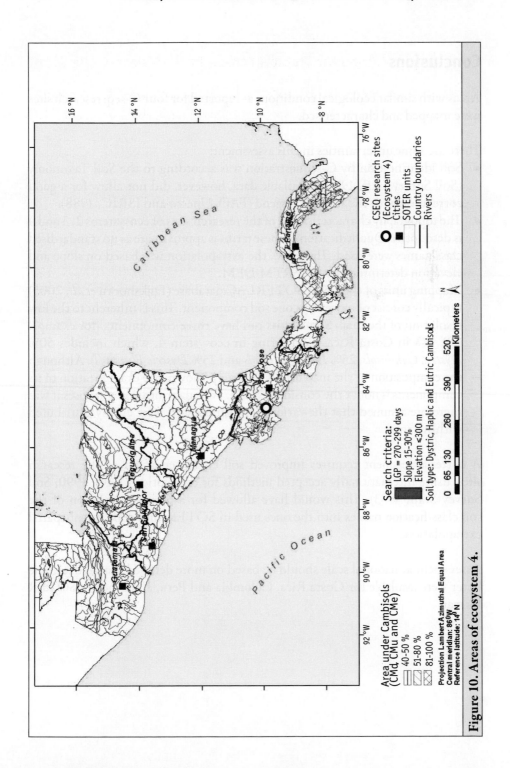

Figure 10. Areas of ecosytems 4.

Conclusions

Areas with similar ecological conditions as reported for four C-seq. research sites were mapped and characterised.

There are some uncertainties in this assessment:
- Soil identification by C-sequestration was according to the Soil Taxonomy (Soil Survey Staff, 1999). Available data, however, did not allow for a good correlation with the Revised Legend (FAO, Unesco and ISRIC, 1988).
- The topography characterisation of the research sites of ecosystems 2, 3 and 4 is descriptive. Quantification of these terms is approximate as no standardised class names were used. Therefore, the extrapolation was based on slope and elevation determined by the SRTM-DEM.
- Mapping units of the 1:5 M SOTERLAC database (Dijkshoorn *et al.*, 2005) typically consist of more than one soil component. This is inherent to the low resolution of the map. Some units can have three components (for example CR 183 in Costa Rica – occurring in ecosystem 4, which includes 50% *Dystric Cambisols*, 25% *Eutric Gleysols* and 25% *Dystric Fluvisols*). Although the composition of the map unit is relatively known, the exact location of its components is not at the considered scale. For quantification purposes it was therefore assumed that the various soil components were evenly distributed within a unit.

A better assessment requires improved soil characterisation of the research sites, using internationally accepted methods for description (FAO, 1990; Soil Survey Staff, 1983). This would have allowed for a better translation of the soil classification names into the ones used in SOTER, and subsequent spatial extrapolation.

Assessment at national scale should be based on more detailed spatial data. The latter were available for Costa Rica, Colombia and Peru, albeit not in SOTER format.

Chapter 10. Conclusions and policy recommendations[1]

J.A. Gobbi, M.C. Amézquita, M. Ibrahim and E. Murgueitio

The C-sequestration project has focused on identifying improved and sustainable pastoral and silvopastoral systems that enhance C accumulation and are economically attractive to farmers located in four Tropical Forest ecosystems of Latin America (Chapter 1). We assessed the accumulation of C in the soil and biomass in different already established pastoral and silvopastoral system in livestock farms in the four ecosystems. We also estimated the financial viability of investing in the implementation of pastoral and silvopastoral systems that enhance C sequestration and explored their contributions to farmers in terms of poverty alleviation by incrementing production and income and by providing additional sources of labour.

In preceding chapters it was established that improved, well-managed pasture and silvopastoral systems have the capacity to sequester significant amounts of C and increase production and income at the same time when applied to recuperate already deforested, degraded pasturelands.

In terms of C sequestration native forests had the highest stocks of accumulated C in the entire system (soil plus biomass), followed, in order, by well-managed improved pasture and silvopastoral systems. Degraded pastures and degraded soil had the lowest stock of accumulated C in the entire system. The largest proportion of accumulated C in pasture and silvopastoral systems was in the soil levels of soil C accumulated by pastures were approaching or higher than in native forests. Improved grass and grass plus legume pastures showed the highest potential to enhance C accumulation. Depending on climate and soil conditions, these systems showed annual rates of C capture of about 2.7 t/ha/yr. When newly established on degraded soils, improved pasture and silvopastoral systems showed soil C sequestration rates in a 3.4-yr experimental period of 1.8 to 7.7 t ha/yr/m.

The stock of C and the accumulation rates of those systems can be further increased if a moderated number of dispersed trees (providing between 15-25%

---

[1] Thanks are due to Professor Manuel Rodríguez for his advice on policy recommendations.

of shade cover) are incorporated in the pasture, either by planting or by natural regeneration. The introduction of dispersed trees in pastures is relevant, besides increasing the stock of C accumulation, for their role in enhancing habitats for biodiversity, as several studies conducted in Tropical America have shown.

Forage banks did not accumulate significant amounts of C, as most of the biomass was exported from the system to feed animals. However, they allow the intensification of production of the farm and potentially release pressure for expanding pastures at the expense of forested areas.

One of the major problems in Latin America is the conversion of forest to unfertilised grass monoculture pastures, which are unsustainable and degrade within a few years. The results of the project showed that both improved legume-grass pastures and silvopastoral systems were sustainable if well managed and stored significant amounts of C in the soil. Developing mosaic land uses of these systems will contribute to increasing diversity and complexity of landscapes while improving productivity of farms and C sequestration. However, the establishment of legume-based pastures, silvopastoral systems and forage banks, requires capital investments that go beyond the financial ability of small farmers and therefore incentive systems should be designed to promote the introduction of these systems.

The economic comparison of the different land uses analysed in this research showed that well-managed, improved pasture and silvopastoral systems present higher levels of production and generate higher net benefits than poorly managed native pasture systems. In addition, the implementation of forage banks substantially increases the use of farm labour, which may increase the demand for labour in the area thus presenting an additional source of employment for family labour. The latter aspect is relevant in areas that offer very limited opportunities for off-farm labour, such as the Colombian Amazon region. Investing in well-managed, improved pasture and silvopastoral systems to recuperate already deforested, degraded pasturelands, was financially profitable and risk free (i.e. the likelihood of negative returns on the investment are quite low). The investments were financially profitable even if decreases in milk and beef production due to tree-grass competition (under high tree density) were considered when modeling the incorporation of dispersed trees in the pasture. Furthermore, pastoral and silvopastoral systems provide a cost-effective opportunity to sequester C, which has the potential to be economically more attractive to farmers than tree plantations, due to almost immediate generation of revenues from milk and beef after the investment. Including revenues from C sequestration in the comparison, estimated according to the tCER scheme devised under the Clean Development Mechanism, showed that the effects of C payments were relatively marginal in terms of making the investment financially attractive. However, for small, cash-poor farmers, a payment for C sequestration may present an important incentive to adopt enhancing C accumulating land use types. Nonetheless, several critical issues associated with the implementation of a payment for C sequestration, such as transaction costs, leakage, additionality and permanence, must be resolved before recommending such an option.

The present study was meant to support the decision-making process regarding the identification of land use options economically attractive to farmers and with capacity to sequestering C. Therefore, and having established that well-managed, introduced pasture and silvopastoral systems meet those criteria, a question leaps to mind: what are the policy implications and recommendations that can be derived from the study findings? An obvious implication is that, since native forests have the highest total C stocks, their destruction should be avoided and their conservation guaranteed in livestock development projects and that degraded pasturelands and degraded soils could be recuperated by implementing well-managed, introduced legume-based pasture and silvopastoral systems as part of a strategy to improve farmers' income and mitigate C emissions. Thus, if such an option is contemplated, two critical aspects to be considered are the capital requirements for the initial investments and the technical assistance to manage these systems. Although there are variations in cost structures among the four areas, these systems present relatively high establishment and operating costs, as is the case in particular of forage banks. Such characteristics determine that policy recommendations to promote the implementation of these systems cannot be proposed across the board, but must take into consideration the different types of farmers present in each area. In practice, small and medium size farms may need additional help and incentives to adopt these systems as their usually limited availability of capital may present a barrier for making the necessary land use changes. In this regard, the stimuli may include loan facilities with extended repayment periods and tax deductions for adopting environmentally friendly land use types. Furthermore, cash-poor farmers need to be presented with strategies to lower the risk of the investment, such as provision of inputs and of technical assistance, since they do not have access to credit. In contrast, 'large farmers'[2] require a different approach, since they generally do not have significant capital constraints for making the initial investment or for obtaining technical assistance. For these farmers, the design of a 'green loan' with a portion of the borrowed money devoted to finance a component associated with production (such as the introduction of improved pastures) and another portion devoted to conservation (such as the incorporation of trees in pastures), may be enough to stimulate the required land use changes. In fact, there is a recognisable trend in the region in which large farms are introducing improved pastures (in particular *Brachiaria* spp.), which implies that these type of improved pastures are proving financially advantageous *vis-à-vis* native pastures since large farms are adopting them without need of external incentives. Therefore, the strategy for large farms must be directed to make the management of their pasturelands more environmentally friendly, such as promoting the incorporation of trees in pastures and the reforestation of areas not suitable for livestock production to favour soil and water protection and biodiversity conservation.

Our study showed that payments for C sequestration in well-managed, introduced pasture and silvopastoral systems are marginal in terms of revenues per unit area, and only landowners with large farms could expect a modest welfare increase through such operations. However, the existence of farms with large tracts of remnant forests, which is common in the Colombian Amazon region, indicates that for these farms it is extremely important to avoid the expansion of pastures at the expense of native forests. In this sense, the option of payments for 'avoided deforestation' (topic not considered by the current operation contemplated in the first stage of the CDM implementation, but under discussion in the UNCCC) may offer a potential opportunity to protect those C stocks in the future.

Implications for global policies on climate change

Results from this project are significant regarding the positions to be adopted by tropical American countries *vis-à-vis* the United Nations Conference on Climate Change and the negotiations carried out through its implementing bodies. The development of a strategy to include improved pasture and silvopastoral systems as well as the conservation of natural forests (avoided deforestation) is required

---

[2] 'Large farmers' refers to farmers that do not have capital constraints, but does not necessarily coincide with farm size.

as eligible projects under the Clean Development Mechanism for the next implementation phase of the Kyoto Protocol. In the case of improved pastoral and silvopastoral systems, this project provides ample evidence that those systems present a valid alternative to increase C accumulation, particularly in soils. In the case of native forests, the results from this investigation underline, once again, the importance native forests play as C stocks compared to other land uses. This fact becomes particularly relevant given the high deforestation rates the region continues to suffer.

Implications for national policies on climate change and land use

Findings from this project are robust enough to encourage governments in tropical America to review their policies that favour reforestation and afforestation as a means to recover degraded pastures and, simultaneously, to mitigate climate change. This study clearly indicated the capacity of improved pasture and silvopastoral systems to recuperate degraded areas and, at the same time, to provide an attractive economic alternative to farmers. Obviously, making this statement does not imply that policies must be adopted to favour improved pasture and silvopastoral systems as the best options to achieve objectives of poverty alleviation and climate change mitigation. The decision on the best alternative (reforestation, afforestation, improved pasture and silvopastoral systems) is, in the end, site specific and must include environmental as well as social considerations, such as norms and regulations on water management, soil and biodiversity protection, poverty levels, labour demands, capital requirements and cultural traditions, among others. A possible application of the results of the present project is in sustainable rural development initiatives that combine compulsory conservation of native forests with promotion of land use changes towards improved pastoral and silvopastoral systems taking into consideration the major advantages of each use as shown in Table 1, where land uses are scored according to their contribution to income generation, rural employment and C sequestration.

Table 1. Contribution of land uses to stock and C sequestration, income, rural employment and food security. The black colour means a greater contribution, the grey colour intermediate and white colour no significant contribution.

Land use	C stock and sequestration	Economic benefit ($)	Rural employed	Food security	ES[1]
Native forest	black	white	white	white	black
Degraded pasture and soil	white	white	white	white	white
Forage banks	white	black	black	black	grey
Improved pasture	black	grey	grey	white	grey
Silvopastoral systems	black	grey	grey	white	grey

[1]Environmental services.

References

Albrecht, A. and Kandji, S. (2003). C sequestration in tropical agroforestry systems. Agricultural Ecosystem and Environment 99, 15- 27.

Amézquita, M.C., Ibrahim, M.A. and Buurman, P. (2004). C-sequestration in pasture, agro-pastoral and silvo-pastoral systems in the American tropical forest ecosystem. In: Mannetje, L't et al., (Eds). The importance of silvopastoral systems in rural livelihoods to provide ecosystem services. Proceedings of the second International Symposium on Silvopastoral Systems. Merida, Yucatan, Mexico pp. 303-306.

Amézquita, M.C., Ibrahim, M.A., Buurman, P. and Amézquita, E. (2005a). C-sequestration in pastures, silvo-pastoral systems and forests in four regions of the Latin American tropics. Journal of Sustainable Forestry 21, 31-49.

Amézquita, M.C., Ibrahim, M.A., Buurman, P. and Amézquita, E. (2005b). C-sequestration in pastures, silvo-pastoral systems and forests in four regions of the Latin American tropics. In: Montagnini, F. (Ed.) Environmental Services of Agroforestry Systems. The Haworth Press, Inc., New York.

Amézquita, M.C., Buurman, P., Murgueitio, E. and Amézquita, E. (2005c). C-sequestration potential of pasture and silvo-pastoral systems in the tropical Andean hillsides. In: Lal, R. et al. (Eds). C-sequestration in Soils of Latin America. The Haworth Press, Inc., New York, pp 267-284.

Andrade, C. (1999). Dinámica productiva de sistemas silvopastoriles con Acacia mangium y Eucalyptus deglupta en el trópico húmedo. Tesis Mag. Sc. CATIE, Turrialba, Costa Rica. 70 pp.

Andrade, H.J. and Ibrahim, M.A. (2003). ¿Cómo monitorear el secuestro de Co en sistemas silvopastoriles? Agroforestería en las Américas 10, 109-116.

Anselin, L. and Griffith, D.A. (1988). Do spatial effects really matter in regression analysis? Papers of the Regional Science Association 65, 11-34.

Ariew, R. (1976). Ockham's Razor: A historical and philosophical analysis of Ockham's principle of parsimony. Champaign-Urbana, University of Illinois.

Banco de la República de Colombia.(2006). Series Estadísticas. Producción, salarios y empleo. Índice de producción real. http://www.banrep.gov.co/series-estadisticas/see_prod_salar.

Binswanger, H. (1989). Brazilian policies that encourage deforestation in the Amazon. Environment Department Working Paper No. 16, World Bank, Washington.

Blair, T. (2004). Speech given by the prime minister on the environment and the 'urgent issue' of climate change. Wednesday September 15, 2004.En Guardian Unlimited, http://politics.guardian.co.uk/green/story/0,9061,1305030,00.html)

References

Bolívar, D., Ibrahim, M.A., Kass, D., Jiménez, F. and Camargo, J.C. (1999). Productividad y calidad forrajera de *Brachiairia humidicola* en monocultivo y en asocio con *Acacia mangium* en un suelo ácido en el trópico Húmedo. Agroforestería en las Américas 6, 48-50.

Brown, S. (1997). Estimating biomass and biomass change of tropical forests: A primer. Rome, IT, FAO. 55p (Forestry Paper 134).

Bruce, J.P., Frome, M., Haites, E., Janzen, H.H., Lal, R. and Paustian, K. (1998). C sequestration in soils. J. Soil Water Conservation 54, 382-389.

Bustamante J., Ibrahim M.A. and Beer J. (1998). Evaluación agronómica de ocho gramíneas mejoradas en un sistema silvopastoril con poró (*Erythrina poeppigiana*) en el trópico húmedo de Turrialba. Agroforesteria en lan Americas. Turrialba 5, 11-16.

Bucher, E., Bouille, D., Navajas, H. and Rodríguez Becerra, M. (2000). Country capacity development needs and priorities. regional report for Latin America and the Caribbean. Global Environment Facility and United Nations Development Program, New York.

Buurman, P., Ibrahim, M.A. and Amézquita, M.C. (2004). Mitigation of greenhouse gas emissions by silvopastoral systems: optimism and facts. In: Mannetje, L't *et al.*, (Eds). The importance of silvopastoral systems in rural livelihoods to provide ecosystem services. Proceedings of the second International Symposium on Silvopastoral Systems. Merida, Yucatan, Mexico, pp. 61-72.

CADMA. (1992). Comisión amazónica de desarrollo y medio ambiente. Amazonia sin Mitos. Banco Interamericano de Desarrollo – Tratado de Cooperación Amazónica – PNUD. 13-46-47 pp.

CATIE and Guelph (2000) Evaluaciones de C en sistemas silvopastoriles. Publicación interna, Proyecto Agroforestería Tropical, CATIE, Diciembre 2000.

Chilès, J-P. and Delfiner, P. (1999). Geostatistics: Modeling spatial uncertainty. John Wiley and Sons, New York.

CIAT, (1976-1999). CIAT's Tropical Pastures Program Annual Reports 1976-1993 and CIAT's Tropical Lowlands Program Annual Reports 1994-1999.

CIAT (1999). Tropical Pastures Program Annual Annual Report 1999.

Cochran, W.G. (1977). Sampling Techniques, Third Edition. John Wiley and Sons, New York.

Coleman K. and Jenkinson, D.S. (1996). RothC-26.3- A model for turnover of C in soil. In: Evaluation of Soil Organic Matter Models Using Existing, Long-Term datasets. D.S. Powlson, P. Smith and J.U. Smith (Eds.). NATO ASI Series I, Vol. 38, Springer-Verlag, Heidelberg, pp. 237-246.

Cressie, N.A.C. (1993). Statistics for spatial data. John Wiley and Sons, New York.

DANE (Departamento Administrativo Nacional de Estadística de Colombia) (1996). Encuesta nacional agropecuaria, resultados 1995. Bogotá, Colombia.

Davis, J.C. (2002). Statistical and data analysis in geology. Third Edition. John Wiley and Sons, New York.

Degryze, S., Six, J., Paustian, K., Morriss, S., Paul, E. and Merckx, R. (2004). Soil organic C pool changes following land-use conversions. Global Change Biology 10, 1120-1132.

Desjardins, T., Barros, E., Sarrazin, M., Girardin, C. and Mariotti, A. (2004). Effects of forest conversion to pasture on soil C content and dynamics in Brazilian Amazonia. Agriculture, Ecosystems and Environment 103, 365-373.

Dijkshoorn, J.A., Huting, J.R.M. and Tempel, P. (2005). Update of the 1:5 million Soil and Terrain Database for Latin America and the Caribbean (SOTERLAC; version 2.0). ISRIC Report 2005/01, ISRIC, Wageningen.

Eamus, D., McGuinness, K. and Burrows, W. (2000). Review of allometric relationships for estimating woody biomass for Queensland, the northern territory and Western Australia. National C accounting system. Technical Report 5a, 56 pp.

Ellert, B.H., Janzen, H.H. and Entz, T, (2002) Assessment of a method to measure temporal change in soil C storage. Soil Science Society of America Journal 66, 1687-1695.

Elliott, E.T., Janzen, H.H., Campbell, C.A., Cole, C.V. and Myers, R.J.K. (1994). Principles of ecosystem analysis and their application to integrated nutrient management and assessment of sustainability. In: R.C. Wood and J. Dumanski (Eds.) Sustainable Land Management for the 21st Century. Volume 2 Proceedings of the International Workshop on Sustainable Land Management for the 21st Century. University of Lethbridge, Lethbridge, Canada, June 20-26, 1993.

Espinel, R. (1994). Sociedad y Economía de Campesinos cafeteros de la cordillera Occidental en el Norte del Valle del Cauca. Factores que inciden en la construcción de sistemas Agrarios. Tesis de Maestría en Desarrollo Sostenible de Sistemas Agrarios – Universidad Javeriana, CIPAV, IMCA, 122 p.

FAO (1990). Guidelines for soil description. FAO, Rome, 70 pp.

FAO (2000). FAOSTAT estadísticas de producción agropecuaria para América Latina y el Caribe. Roma, Italia.

FAO (2002). Food balance sheets. Rome, Italy.

FAO ISRIC, UNEP and CIP (1998). Soil and Terrain Database for Latin America and the Caribbean – SOTERLAC (version1.0). Land and Water Digital Media Series 5, Food and Agriculture Organization (FAO) of the United Nations, the International Soil Reference and Information Centre (ISRIC), the United Nations Environment Programme (UNEP) and the International Potato Centre (CIP), Rome.

FAO, UNESCO and ISRIC (1988). Revised Legend, Soil Map of the World. World Soil Resources Reports 60, FAO, Rome.

FAO-IIASA 2000. Global Agro-Ecological Zones, FAO-IIASA (http://www.iiasa.ac.at/Research/LUC/GAEZ/index.htm).

References

Fearnside, F.M. and Barbosa, R.I. (1998). Soil C changes from conversion of forest to pasture in Brazilian Amazonia. Forest Ecology and Management 108, 147-166.

Fearnside, M. (1979). Cattle yield prediction for the trans-Amazon highway of Brazil. Interciencia 4, 220-225.

Fisher, M.J., Braz, S.P., Dos Santos, R.S.M., Urquiaga, S., Alves, B.J.R. and Boddey, R.M. (2007). Another dimension to grazing systems: Soil C. Tropical Grasslands 41, 65-83.

Fisher, M.J. and Thomas, R.J. (2004). Implications of Land-use change to introduced pastures on C stocks in the central lowlands of tropical South America. Environment, Development and Sustainability 6, 111-131.

Fisher, M.J., Rao, I.M., Ayarza, M.A., Lascano, C.E., Sanz, J.I., Thomas, R., Thomas, J. and Vera, R.R. (1994). C storage by introduced deep-rooted grasses in the South American savannas. Nature 371, 236-238.

Flannery, T. (2006). The Weather Makers: How Man Is Changing the Climate and What It Means for Life on Earth. Atlantic Monthly Press, New York.

Fortin, M-J. and Dale, M. (2005). Spatial analysis, a guide for ecologists. xiii + 365 p. Cambridge University Press, Cambridge.

Gibbons, J., Olkin, I. and Sobel, M. (1977). Selecting and ordering populations: A new statistical methodology. John Wiley, New York.

Gittinger, J.P. (1982). Economic analysis of agricultural projects. Johns Hopkins University.

Glendining, M.J. and Poulton, P.R. (1996). Interpretation difficulties with long-term. In: Evaluation of soil organic matter models using existing, long-term datasets, D.S. Powlson, P. Smith and J.U. Smith (Eds.) NATO ASI Series I, Vol. 38, Springer-Verlag, Heidelberg, pp. 99-109.

Gore, A. (2006). An inconvenient truth: the planetary emergency of global warming and what we can do about it. Melcher Media/Rodale, New York.

Guo, L.B. and Gifford, R.M. (2002). Soil C stocks and Land-use change: a meta analysis. Global Change Biology 8, 345-360.

Hansen, J. (2006). The Threat to the Planet. The New York Review of Books, Volume 53, No. 12, July 13, 2006.

Hanson, J.D., Schaffer, M.J., and Ahuja, L.R. (2001). Simulating rangeland production and C sequestration. In: R.F. Follett, J.M. Kimble and R. Lal (Eds.): The potential of U.S. grazing lands to sequester C and mitigate the greenhouse effect. CRC Press, Boca Raton London New York Washington D.C., pp. 345-370.

Hecht, S. (1992). The logics of livestock and deforestation. In: Development or destruction. The conversion of tropical forest to pasture in Latin America. Westview Press/Boulder. San Francisco. 7 pp.

Hénin, S. and Dupuis, M.(1945). Essai de bilan de la matière organique du sol. Annales Agronomiques 15, 17-29.

Heckrath, G., Djurhuus, J., Quine, T.A., Van Oost, K., Govers, G. and Zhang, Y. (2005). Tillage erosion and its effect on soil properties and crop yield in Denmark. Journal of Environmental Quality 34, 312-332.

Hochberg, Y. and Tamhane, A.C. (1987). Multiple comparison procedures. Wiley Series in Probability and Mathematical Statistics. 450pp. New York: John Wiley.

Hogg, R.V., McKean, J.W. and Craig, A.T. (2005). Introduction to mathematical statistics. Sixth Edition. Pearson Prentice Hall, Upper Saddle River, NJ.

Holguin, V, Ibrahim, M.A., Mora, J., and Rojas, A. (2003). Caracterización de sistemas de manejo nutricional en ganaderías de doble propósito de la región Pacifico Central de Costa Rica. Agroforestería en las Américas 10, 40-46.

Ibrahim, M.A., Franco, M., Pezo, D.A., Camero, A. and Araya, J.L. (2001). Promoting intake of *Cratylia argentea* as a dry season supplement for cattle grazing *Hyparrhenia rufa* in the subhumid tropics. Agroforestry Systems 51, 167-175.

Ibrahim, M.A. and 't Mannetje, L. (1998). Compatibility, persistence and productivity of grass-legume mixtures in the humid tropics of Costa Rica. 1. Dry matter yield, nitrogen yield and botanical composition Tropical Grasslands 32, 96-104.

INEC (Instituto Nacional de Estadística y Censos CR) (2001). IX Censo Nacional de Población y V de Vivienda, Resultados Generales. San José, Costa Rica, 80 p.

IPCC (1999). Climate change and its linkages with development, equity, and sustainability. Proceedings of the IPCC expert meeting held in Colombo, Sri Lanka, 27-29 April 1999.

Isaaks, E.H. and Srivastava, R. M. (1989). An introduction to applied geostatistics. xix + 561 pp. Oxford University Press, New York, Oxford.

IVITA (1981). Informe anual. Lima, Peru, 75 pp.

Jackson, J. and Ash, A. (1998). Tree-grass relationships in open eucalypt woodlands of northern Australian: influence of trees on pasture productivity, forage quality and species distribution. Agroforestry Systems 40, 159-176.

Kaimowitz, D. (1996) Livestock and deforestation in Central America in the 1980s and 1990s: A policy perspective. Center for International Forestry Research (CIFOR), Special Publication, Jakarta, 88.

Kaimowitz, D. (2000). Useful myths and intractable truths: the politics of the link between forests and water in Central America. San Jose, Costa Rica, Center for International Forestry Research (CIFOR). San José de Costa Rica.

Kanevski, M. and Maignan, M. (2004). Analysis and modelling of spatial environmental data. xi + 288 p. EPFL Press, Marcel Dekker, Inc., Lausanne, Switzerland.

Kolbert, E. (2006). Field Notes from a Catastrophe: Man, Nature, and Climate Change. Bloomsbury, New York.

Kravchenko, A.N. (2003). Influence of spatial structure on accuracy of interpolation methods. Soil Science 67, 1564-1571.

References

Kravchenko, A.N., Robertson, G.P. Snap, S.S. and Smucker, A.J.M. (2006). Using Information about spatial variability to improve estimates of total soil C. Agronomy Journal 98, 823-829.

Krugman, P. (2006). Enemy of the planet. In New York Times, April 17, 2006.

Lal, R. (2004). Soil C sequestration impacts on global climate change and food security Science 304, 1623-1627.

Lascano, C.E., Amézquita, M.C., Avila, P. and Ramírez, G. (2001) Sources of Variation in Milk Production and Composition of Dual-Purpose Cows under Sequential Grazing. In: Amézquita, M.C. Biometrical Applications in Tropical Pastures and Agro-pastoral Research, Wageningen University, 2001. Adapted from the article published in Spanish in: Lascano, C.E. and Holmann, F. (Eds.), 1997. Metodologías de Investigación en Fincas con Sistemas de Producción Animal de Doble Propósito. CIAT-TROPILECHE. Book Series, 258 pp.

Lehmann, E.L. and Romano, J.P. (2006). Testing statistical hypotheses. Texts in Statistics. Springer, Heidelberg. 786 pp.

Loetsch, F., Zöhrer, F.and Haller, K.E. (1973). Forest inventory, Volume 2. Munich: BLV Verlagsgesellschaft.

López, A., Schlonvoigt A., Ibrahim M., Kleinn C and Kanninen M. 1999. Cuantificación del Co almacenado en el suelo de un sistema silvopastoril en la zona Atlántica de Costa Rica. Revista Agroforestería en las Américas 6, 51-53.

Louman, B. (2002). Inventarios en bosques secundarios. In: Inventarios forestales para bosques latifoliados en América Central. L. Orozco and C. Brumér (Eds.). Turrialba, Costa Rica CATIE. 264 pp.

Lugo, A.E., Sánchez, M.J. and Brown, S. (1986). Land use and organic C content of some subtropical soils. Plant and Soil 96, 185-196.

MacDicken, K. (1997). A guide to monitoring C storage in forest and agroforestry projects. forest C monitoring program. Winrock International Institute for Agricultural Development.

MAG (Ministerio de Agricultura y Ganadería). (2001). Censo Nacional Agropecuario 2000. San José, Costa Rica.

Márquez, L., Roy, A. and Castellanos, E. (2000). Elementos técnicos para inventarios de Co CO_2 en uso del suelo. Guatemala, Guatemala, Fundación Solar. 36 pp.

McBratney, A.B. and Pringle, M.J. (1998). Estimating average and proportional variograms of soil properties and their potential use in precision agriculture. Precision Agriculture 1, 125-152.

Minami, K., Goudriaan, J., Lantinga, E.A. and Kimura, T. (1993). Significance of grasslands in emission and absorption of greenhouse gases. Proceedings of the XVIIth International Grassland Congress 2, 1231-1238.

Mora C.V. (2001). Fijación, emisión y balance de gases de efecto invernadero en pasturas en monocultivo y en sistemas silvopastoriles de fincas lecheras intensivas de las zonas altas de Costa Rica. Tesis Mag. Sc. CATIE, Turrialba, Costa Rica. 92 p.

Müller, M.M.L., Guimaraes, M.F., Desjardins, T. and Mitja, D. (2004). The relationship between pasture degradation and soil properties in the Brazilian Amazon: a case study. Agriculture, Ecosystems and Environment 103, 279-288.

Muñoz, J. (2004). Alternativas de uso del suelo en terrazas aluviales de la Amazonia Colombiana. In: Ramírez, B.L., Estrada, C.A., Rodríguez, J.G., Muñoz, J. and Guayara, A. (Eds.). Aporte al conocimiento y sostenibilidad del agroecosistema intervenido de la Amazonia colombiana. Centro de Investigación y Desarrollo de Sistemas Sostenibles de Producción Amazónica, Cidespa, Universidad de la Amazonia. Impresora Feriva. Florencia (Caquetá, Colombia). pp: 141-176.

Neil, C., Melillo, J.M., Seudler, P.A. and Cerrl, C.C. (1997). Soil C and nitrogen stocks following forest clearing for pasture in the southwestern Brazilian Amazon. Ecological Applications 7, 1216-1225.

Nyberg, G. and Högberg, P. (1995). Effects of young agroforestry trees on soils in on-farm situations in eastern Kenya. Agroforestry Systems 32, 45-52.

Odeh, I.O.A., McBratney, A.B. and Chittleborough, D.J. (1990). Design of optimal sample spacings for mapping soil using fuzzy-k-means and regionalized variable theory. Geoderma 47, 93-122.

Olea, R.A. (2003). Geostatistics for engineers and earth Scientist. Kluwer Academic Publishers, Boston, Dordrecht, London.

Overman, J.P.M., Witte, H.J.L. and Saldarriaga, J.G. (1994). Evaluation of regression models for above-ground biomass determination in Amazon rainforest. Journal of Tropical Ecology 10, 207-218.

Overmars, K.P, De Koning G.H.J. and Veldkamp, A. (2003). Spatial autocorrelation in multi-scale land use models. Ecological Modelling 164, 257-270.

Pachico, D., Ashby, J. and Sanint, L.R. (1994). Natural resource and agricultural prospects for hillsides of Latin America. Paper prepared for IFPRI 2020-Vision. Workshop Washington D.C., 7-10 November. CIAT's Hillsides Program Annual Report 1993-1994, pp. 283-321, CIAT, Cali, Colombia.

Parresol, B. (1999). Assessing tree and stand biomass: a review with examples and critical comparsions. Forest Science 45, 573-593.

Pandey, D.N. (2000). C sequestration in agroforestry systems. Climate Policy 2, 367-377.

Parton, W.J., Sanford Jr., R.L., Sanchez, P.A. and Stewart, J.W.B. (1989). Modeling soil organic matter dynamics in tropical soils. In: B. Bohlool, D. Coleman and G. Uehara (Eds.). Dynamics of soil organic matter in tropical ecosystems. University of Hawaii Press, Honolulu. pp. 153-171

Parton, W.J. (1996). The CENTURY model. In: Powlson, D. S. Smith, P.and Smith J. U. (Eds.). Evaluation of soil organic matter models using existing long-term datasets. pp. 283-291.Springer-Verlag, Berlin.

Parton, W.J., Scurlock, J.M.O., Ojima, D.S., Gilmanov, T.G., Scholes, R.J., Schimel, D.S., Kirchner, T., Menaut, J.C., Seastedt, T., Garcia, M., Apinan Kamnalrut, E. and Kinyamario, J.I. (1993). Observations and modeling of biomass and soil organic matter dynamics for the grassland biome worldwide. Global Biogeochemical. Cycles 7, 785-809.

Parton, W.J., Stewart, J.W.B. and Cole, C.V. (1987). Dynamics of C, N, S, and P in grassland soils: A model. Biogeochemistry 5, 109-131.

Pfaff, A.S.P., Kerr, S., Cavatassi, R., Davis,B, Lipper, L., Sanchez, A. and Timmins, J. (2004) Effects of poverty on deforestation: Distinguishing behavior from location. ESA Working Paper No. 04-19.

Post, W.M. and Kwon, K.C. (2000). Soil C sequestration and land-use change: processes and potential. Global Change Biology 6, 317-327.

Powers, J.S., and Veldkamp, E. (2005). Regional variation in soil C and δ13C in paired forests and pasture of Northeastern Costa Rica. Biogeochemistry 72, 315-336.

Powlson, D.S. (1996). Why evaluate soil organic matter models? In: Powlson, D.S., Smith, P. and Smith, J.U. (Eds.). Evaluation of soil organic matter models using existing long-term datasets. Springer-Verlag, Berlin, pp. 3-11.

Powlson, D.S., Smith, P. and Smith J.U. (Eds.) (1996). Evaluation of soil organic matter models. NATO Advanced Science Institute Series, Springer-Verlag, Berlin, Heidelberg, 429 pp.

Ramírez, B. L. (2004). Diagnóstico ambiental y alternativas de desarrollo sostenible en fincas ganaderas establecidas en la Amazonia colombiana. In: Ramírez, B.L., Estrada, C. A., Rodríguez, J.G., Muñoz, J. and Guayara, A. (Eds.) Aporte al conocimiento y sostenibilidad del agroecosistema intervenido de la Amazonia colombiana. Centro de Investigación y Desarrollo de Sistemas Sostenibles de Producción Amazónica, Cidespa, Universidad de la Amazonia. Impresora Feriva. Florencia (Caquetá, Colombia). pp. 17-57.

Restrepo, C., Ibrahim M.A. and Harvey, C.A. (2004). Relations between tree cover in pasturelands and animal production in cattle farms in the dry tropic area of Canas, Costa Rica. In: Mannetje, L't et al., (Eds). The importance of silvopastoral systems in rural livelihoods to provide ecosystem services. Proceedings of the Second International Symposium on Silvopastoral Systems. Merida, Yucatan, Mexico, pp. 193-196.

Rhoades, C.C., Eckert, G.E and Coleman, D.C. (1998). Effect of pasture trees on soil nitrogen and organic matter: implications for tropical montane forest restoration. Restoration Ecology 6, 262-270.

Rodríguez Becerra, M and Espinoza, G (2002). Gestión ambiental en América Latina y el Caribe: evolución, tendencias y principales prácticas, Inter American Development Bank, Washington, D.C.

Roscoe, R., Buurman, P., Velthorst, E.J. and Vasconcellos, C.A. (2001). Soil organic matter dynamics in density and particle size fractions as revealed by the $^{13}C/^{12}C$ isotopic ratio in a Cerrado's oxisol. Geoderma 104, 185-202.

Rosero, M. (2000). Scaling-up through complexity. The role of everyday rural life and informal social networks in the dissemination of knowledge about enviromental-friendly agricultural technologies. The case of Bellavista Community, Colombia. MSc, Thesis, Wageningen University and Research Centre. 209 p.

Saltelli, A., Tarantola, S., Campolongo, F. and Ratto, M. (2004). Sensitivity analysis in practice. xi + 219p. John Wiley and Sons, Ltd. Chichester, West Sussex, England.

Sánchez, P., Bandy, D., Villachica, J. and Nicholaides, J. (1982). Amazon basin soils: Management for continuous crop production. Science 216, 821-827.

Shukla, M.K., Slater, B.K., Lal, R. and Cepuder, P. (2004). Spatial variability of soil Properties and potential management classification of a chernozemic field in lower Austria. Soil Science 169, 852-860.

Segura, M. and Kanninen, M. (2002). Inventario para estimar Co en ecosistemas forestales tropicales. In: L. Orozco and C. Brumér (Eds.) Inventarios forestales para bosques latifoliadas en América Central. CATIE. 264 p.

Segura, M., Kanninen, M. and Suárez, D. (2006). Allometric models for estimating aboveground biomass of shade trees and coffee bushes grown together. Agroforestry Systems 68, 143-150.

Serrao, E., Falesi, J., Veiga, A. and Texeira, J. (1978). Productivity of cultivated pastures on low fertility soils of the Amazon Basin. EMBRAPA, Belém do Pará, Brazil.

Shirato, Y., Paisancharoen, K., Sangtong, P., Nakviro, C. Yokozawa, M.,Matsumoto, N. (2005). Testing the Rothamsted C model against data from long-term experiments on upland soils in Thailand. European Journal of Soil Science 56, 179-188.

Silver, W.L., Ostertag, R. and Lugo, A.E. (2000). The potential for C sequestration of abandoned tropical agricultural and pasture lands. Restoration Ecology 8, 394-407.

Sioli, H. (1980). Foreseeable consequences of actual development schemes and alternative ideas. In: Barbira-Scazzochio, F. (Ed.). Land, people and planning in contemporary Amazonia. Centre of Latin American Studies, Cambridge University Press, New York.

Smith, P., Powlson, D.S. Smith, J.U. and Elliott, E.T. (1997). Evaluation and comparison of soil organic matter models using datasets from seven long-term experiments. Geoderma 81, 1-225.

References

Smith, J., Smith, P. and Addiscott, T. (1996). Quantitative methods to evaluate and compare soil organic matter models. In: Powlson, D.S. Smith, P. and Smith, J.U. (Eds.), Evaluation of soil organic matter models using existing long-term datasets. Springer-Verlag, Berlin, pp. 181-199.

Soil Survey Analytical Continuum (1996). Soil survey laboratory methods manual. Soil survey investigations report 42. US Department of Agriculture, Washington.

Soil Survey Staff (1983). Soil survey manual (revised edition). Agricultural Handbook 18, USDA, Washington.

Soil Survey Staff (1999). Soil Taxonomy, A basic system of soil classification for making and interpreting soil surveys. Agricultural Handbook, 438. U.S. Government Printer, Washington DC.

Souza de Abreu, M.H., Ibrahim, M. A. and Silva, J.C. (1999). Arboles en pastizales y su influencia en la producción de pasto y leche. In: Congreso Latinoamericano sobre Agroforestería para la Producción Animal Sostenible. Memorias. CIPAV, Cali, Colombia

Szott, L., Ibrahim, M.A. and Beer, J., (2000). The hamburger connection hangover: cattle pasture land degradation and alternative land use in Central America. CATIE-DANIDA-GTZ.

Tarré, R., Macedo, R., Cantarutti, R.B., de P. Rezende, C., Pereira, J.M., Ferreira, E., Alves, B.J.R., Urquiaga, S. and Boddey, R.M. (2001). The effect of the presences of a forage legume on nitrogen and C levels in soils under *Brachiaria* pastures in the Atlantic region of South of Bahia, Brazil. Plant and Soil 234, 16-26.

TCA. (1991). Reunión de la comisión especial de medio ambiente de la Amazonia. (CEMA). Belém, Brasil.

Terra, J.A., Shaw, J.N., Reeves, D.W. Raper, R.L., Van Santen, E. and Mask P.L. (2004). Soil C relationships with terrain attributes, electrical conductivity, and a soil survey in a coastal plain landscape. Soil Science 169, 819-831.

UNFCCC COP 6 (2000). United Nations Framework Convention on Climate Change, of the Parties at its sixth session, November 13-24, The Hague, The Netherlands.

UNFCCC COP 7 (2001). United Nations Framework Convention on Climate Change, Conference of the Parties at its seventh session, October 29-November 9, 2001, Marrakech, Morocco.

USDA (1996). Soil survey laboratory methods manual. Soil Survey Investigations Report No. 42, Version 3, United States Department of Agriculture, Washington D. C., U.S.A., 693 pp.

USDA (1999). Soil taxonomy: A basic system of soil classification for making and interpreting soil surveys. Agricultural Handbook, 438. U.S. Government Printer, Washington D.C.

USGS (2003). Shuttle radar topographic mission digital elevation model (http://srtm.usgs.gov/).

Velásquez, J., Cipagauta, M., Gómez, J. and Tapia, M. (1999). Avances de la caracterización estática de las empresas ganaderas de doble propósito del piedemonte amazónico colombiano. In: Seminario Técnico, Tecnología para la producción de leche y carne en regiones del trópico bajo colombiano: Orinoquía y Amazónia. Villavicencio, Colombia.

Van Keulen, H. (2001). (Tropical) soil organic matter modelling: problems and prospects. Nutrient Cycling in Agroecosystems 61, 33-39.

Van Putten, B. and Knippers, T.S. (2007). Design and analysis of soil science experiments. (in press)

Veldkamp, E. (1993). Soil organic C dynamics in pastures established after deforestation in the humid tropics of Costa Rica. Ph.D. thesis. Wageningen Agricultural University. The Netherlands. 112p.

Veldkamp, E. (1994). Organic C turnover in three tropical soils under pasture after deforestation. Soil Science Society of America Journal 58, 175-180.

Wadsworth, F.H. (1997). Forest production for Tropical America. United States Department of Agriculture, Agriculture Handbook 710. 563 pp.

Walkley, A. and Black, I.A. (1934). An examination of the Degtjareff method for determining soil organic matter, and proposed modification of the chromic acid titration method. Soil Science 37, 29-38.

Wasserman, L. (2007). All of Nonparametric Statistics. XII + 268 pp. Springer Texts in Statistics. Springer, Heidelberg.

Watson, R.T., Noble, I.R., Bolin, B, Ravindranath, N.H., Verardo, D.J. and Dokken, D.J. (2000). Land Use, Land-Use Change, and Forestry. 375 pp. Cambridge University Press, Cambridge, UK.

Acknowledgements

We express our gratitude to The Netherlands Ministry of Development Cooperation in The Hague and to The Netherlands Embassy in Bogotá, Colombia, for having made possible this 5-year project whose scientific results are reported in this book. We specially thank Mr. Vincent van Zeijst, Mr. Jaques Remmerswaal and Mr. Maurice van Beers, of The Netherlands Embassy in Bogotá, Colombia, for their motivating support.

We thank the partner institutions, consultants, project members, field assistants and co-researchers, students who conducted thesis work in the project, and particularly the farmers on whose farms the research was conducted. All participants are listed on he following pages.

We also thank other Universities whose students conducted thesis work in our project. They are: Universidad Javeriana, Bogotá, Colombia; Universidad Santo Tomás, Cali, Colombia, Montpellier University, France; and Ghent University, Belgium.

M.C. Amézquita
B.L. Ramirez
M.A. Ibrahim

Participating organizations, scientists, assistants, students and farmers

Project organisations

CIPAV, Centre for Research on Sustainable Agricultural Production Systems, Cali, Colombia.
UNIVERSIDAD DE LA AMAZONIA, Florencia, Colombia.
CIAT, International Centre for Tropical Agriculture, Cali, Colombia.
CATIE, Centre for Tropical Agriculture and Education, Turrialba, Colsta Rica.
WUR, Wageningen University and Research Centre, Wageningen, The Netherlands
ISRIC, World Soil Information, Wageningen, The Netherlands

Project participants

Project executive committee

Dr. María Cristina Amézquita
Ph.D. Production Ecology and Resource Conservation
Project Scientific Director
m.amezquita@cgiar.org

Mr. Enrique Murgueitio
CIPAV's Executive Director
Project Administrative and Financial Director
enriquem@cipav.org.co

Dr. Bertha Leonor Ramírez
Universidad de la Amazonía
Ph.D. Agroforestry Systems
Humid tropical forest, Amazonia, Colombia
belerapa@lhotmail.com

Dr. Edgar Amézquita
CIAT
Ph.D. Soil Scientist
e.amezquita@cgiar.org

Dr. Muhammad A. Ibrahim
CATIE
Ph.D. Agronomy
Humid and sub-humid tropical forest, Costa Rica
mibrahim@catie.ac.cr

Dr. Bram van Putten
Wageningen University and Research Centre
Ph.D. Mathematics
Bram.vanPutten@wur.nl

Dr. Peter Buurman
Wageningen University and Research Centre
Ph.D. Soil Chemistry and Dynamics
Peter.Buurman@wur.nl

Consultants

Emeritus Professor Leendert 't Mannetje
Wageningen University and Research Centre
ltmannet@quicknet.nl

Dr. M.J. Fisher
CIAT
m.fisher@cgiar.org

Professor Manuel Rodríguez
Universidad de los Andes
Bogotá, Colombia
and
International Consultant
Environmental Policy Issues

Other project members

Field research – Hillsides ecosystem (Colombia)

Ms. María Elena Gómez
CIPAV
M.Sc. Agronomist
mariae@cipav.org.co

Ms. Piedad Cuellar
CIPAV
M.Sc. Participatory Researcher
piedad@cipav.org.co
B.Sc. thesis students:

Ms. Lucero Gómez
Universidad Santo Tomás
Cali, Colombia

Ms. Laura Cabrera
Universidad Javeriana
Bogotá, Colombia

Mr. Gabriel González
Universidad Javeriana
Bogotá, Colombia

Co-researchers:

Mr. Julián Giraldo
El Ciprés Natural Reserve
El Dovio, Colombia

Ms. Mayerly Guzmán
El Cambio and Villa Victoria Faros,
Dagua, Colombia

Field research – Humid and sub-humid tropical forest, Costa Rica

Mr. Tangaxuhan Llanderal
Ph.D. (cand.) Agroforestry
tllander@catie.ac.cr

Mr. Alexander Navas
CATIE
Agronomist
anavas@catie.ac.cr

Mr. Francisco Casasola
CATIE
Agronomist
fcasasol@catie.ac.cr

Mr. Fabio Mora
CATIE
Agronomist
fmora@catie.ac.cr

Mr. Mario Chacón
CATIE
Agronomist
mchacon@catie.ac.cr

Field research – Humid tropical forest, Amazonia, Colombia

Dr. Jaime Enrique Velásquez
Ph.D. Agronomy

Dr. Jader Muñoz
Ph.D. Geology

B.Sc. students:

Mrs. Jaime Andrés Montilla
Juan Carlos Suárez
Wilmar Yovany Bacón
Ms. Edna Rocío Castañeda

Environmental Economist

Dr. José Gobbi
CATIE
Ph.D. Economics
(Present Address INTA, Argentina)
jgobbi@correo.inta.gov.ar

Soil Scientist

Vincent van Engelen
ISRIC
Wageningen, The Netherlands
Vincent.vanEngelen@wur.nl

Statistician/database analyst

Mr. Héctor Fabio Ramírez
Statistician

Soil and biomass C sampling

Mr. Hernán Giraldo
Agronomist
giraldogo@hotmail.com

Executive Assistant

Mr. Francisco Ruiz
M.Sc. Economics
F.ruiz@cgiar.org

International Students

Mr. Octavio Mosquera
CIAT
Ph.D. student,
Wageningen University, The Netherlands

Ms. Annelies Verlee
M.Sc. student
Ghent University, Belgium

Ms. María Camila Rebolledo
B.Sc. student
Montpellier University, France

Farmers

Mr. Tiberio Giraldo
El Ciprés Natural Reserve
El Dovio, Colombia

Ms. Graciela Guzmán
Villa Victoria Farm
Dagua, Colombia

Ms. Ernestina Alvarez
El Cambio Farm
Dagua, Colombia

Mr. Gerardo Silva
La Guajira Farm
Florencia, Amazonia, Colombia

Escobar family
Pekín Farm
Florencia, Amazonia, Colombia

Mr. Rodrigo Silva
La Palma Farm
Florencia, Amazonia, Colombia

Mr. Antonio López
El Chaparrón
Esparza/Puntarenas, Costa Rica

Index

A
Acacia mangium – 31, 34, 55, 66
Acrocomia aculeata – 34
afforestation – 24
agroforestry systems – 66
allometric equations – 44, 46
Amazon – 32, 33, 36, 39
Andean Hillsides – 32, 36, 38
arable crops – 65
Arachis pintoi – 31

B
beef production systems – 124
biodigestors – 122
biodiversity – 194
bulk density – 69

C
carbon sequestration – 29, 31
 – by newly established improved
   land use systems – 57
 – by the forage bank – 61
 – in long-established land use
   systems – 49
carbon stocks
 – distribution in the soil profile
   – 62
 – in soil – 51, 55
 – in tree biomass – 55
 – in vegetation – 51
 – modelling – 146
cattle
 – beef – 133
 – calving interval – 130
 – cut-and-carry – 124, 130
 – dual-purpose – 30, 33, 113, 124,
   133

 – extensive production – 129
 – farming systems – 129
 – management – 115, 117, 133
 – parturition rate – 130
 – production – 126
CDM – *See:* Clean Development
   Mechanism
Cedrella odorata – 34
char – 54
Clean Development Mechanism
   (CDM) – 21, 24
climate change – 22, 23
conservation – 24
Cordia alliodora – 34
Costa Rica – 33, 37, 40
 – characteristics of the region
   – 132
Cratylia argentea – 35

D
data consistency – 69
deforestation – 24
 – avoided – 196
dispersed trees – 137, 194

E
economic analysis – 127
ecosystems, characterisation – 173
Enterelobium ciclocarpum – 34
Erythrina spp. – 66
establishment costs – 116
 – for improved pasture – 136
 – silvopastoral systems – 130
Eucalyptus deglupta – 66
extrapolation
 – criteria – 178
 – method – 177